A WATER RESOURCES TECHNICAL PUBLICATION

HYDRAULIC LABORATORY TECHNIQUES

A guide for applying engineering knowledge to hydraulic studies based on 50 years of research and testing experience.

UNITED STATES DEPARTMENT OF THE INTERIOR

Bureau of Reclamation

Denver, Colorado

1980

M
SI METRIC

UNITED STATES GOVERNMENT PRINTING OFFICE
DENVER: 1980

REPRINTED 1987
REPRINTED 1989

For sale by the Superintendent of Documents, U.S. Government Printing Office, Washington, D.C. 20402, and the Bureau of Reclamation, Denver Office, Denver Federal Center, P O Box 25007, Denver, Colorado 80225-0007, Attention: D-7923A

PREFACE

The 1980 publication of this book marks the golden anniversary of the Bureau of Reclamation's first model testing in 1930. These past 50 years of research and testing should serve well as a foundation for future accomplishments in this specialized activity.

Hydraulic Laboratory Techniques was prepared as an aid in applying engineering knowledge to hydraulic studies. Consequently, emphasis is placed on basic principles of similitude; techniques of model design, construction, and operation; equipment; and field studies. The principles set forth have been used successfully by the Hydraulics Branch of the Bureau of Reclamation.

The presentation stresses the need for understanding the purpose of hydraulic structures and equipment, and thus gives prominence to the conviction that the laboratory specialist must appreciate the criteria and limitations that affect the designer's work. Familiarity with design considerations is also essential for properly conducting field studies.

This publication is the outgrowth of an earlier Bureau writing–Engineering Monograph No. 18, *Hydraulic Laboratory Practice*, issued in March 1953 and revised in July 1958. *Hydraulic Laboratory Techniques* represents the combined writing efforts of members of the Hydraulics Branch of the Division of Research and David L. Goodman, retired technical editor, Engineering and Research Center, Denver, Colorado. J. C. Schuster, Head, Hydraulics Research Section, Hydraulics Branch, was the principal compiler and editor. Most of the photographs are by W. M. Batts. Many of the illustrations were prepared by the Division of Design, Drafting Branch, from numerous technical reports published by the Hydraulics Branch. Final review and preparation of the manuscript for publication were performed by R. D. Mohrbacher, General Engineer, Technical Publications Branch.

The authors and editors express their thanks to the staff of the Photocomposition Unit and the printing and graphics specialists, who worked many long hours in preparing the edited manuscript for printing.

CONTENTS

Chapter 1 – Laboratory Equipment

Chapter 2 – Instrumentation

Chapter 3 – Model Design

Chapter 4 – Laboratory Studies

Chapter 5 – Free-Surface Flow

A. INTRODUCTION

Chapter 9 – Control of Flow

A. GENERAL

B. CLASSIFICATION OF CONTROLS

Chapter 10 – Pumps, Turbines, and Energy Absorbers

Chapter 11 – Field Studies

A. GENERAL

B. MEASUREMENTS

TABLES

FIGURES

INTRODUCTION

The Bureau of Reclamation began its hydraulic model testing in August 1930 when a small staff of engineers, carpenters, and laborers began work in the Hydraulic Laboratory of the Colorado Agricultural Experiment Station (now Colorado State University), Fort Collins, Colorado. Other Bureau hydraulic laboratories were established and operated in various locations as a particular need arose.

During the clement seasons of 1931 to 1936, an outdoor laboratory was operated on the South Canal of the Uncompahgre Project near Montrose, Colorado, where a discharge of 5.7 cubic meters of water per second under a head of 15 meters was available. This laboratory was used to test large models of the side-channel spillways for Hoover Dam, a complete model of the Imperial Dam and its appurtenant works, and a model of the Grand Coulee Dam spillway bucket.

Concurrently, a hydraulics laboratory was established in the basement of the Old Custom House in Denver, Colorado, where many of the smaller structures designed by the Bureau were studied during the period 1934 to 1937. In 1937, the equipment in the Old Custom House was moved to a small but conveniently located laboratory constructed as an addition to the New Custom House. In 1935, the Fort Collins laboratory building was expanded to about four times its original size to meet the ever-increasing load of assignments. But, the laboratory in the New Custom House gradually absorbed personnel from the Fort Collins laboratory, and by the fall of 1938 the Bureau had completely discontinued its operation at the Fort Collins laboratory.

In 1939, laboratory facilities were installed in the Arizona canyon wall outlet house at Hoover Dam to utilize the available discharge of 5.7 cubic meters per second under a head of 107 meters. This laboratory was operated for several months in 1940 and 1941, and again in 1945, to complement model studies made in the Denver laboratory at lower heads and discharges.

In the latter part of 1946, the Hydraulic Laboratory was moved from the New Custom House to its present location in building 56 at the Denver Federal Center. The laboratory is part of the Bureau of Reclamation's E&R Center (Engineering and Research Center).

Most of the investigations by the Hydraulics Branch are undertaken at the request of the design branches or other offices at the E&R Center. Other studies are requested by project or regional offices. In addition to Bureau of Reclamation projects, Hydraulic Laboratory personnel become involved in studies for or in cooperation with other Government agencies, universities, and foreign countries.

The Hydraulics Branch is staffed and equipped to study, by analytical methods, scale models, and field investigations, problems arising in connection with many types of projects. In the civil engineering field, the studies may include spillways, outlet works, canal chutes, drops, irrigation distribution systems, diversion works, sediment control works, river channels, ground-water flow, subsurface drainage of irrigated lands, reaeration, and stratified flow in reservoirs. In the field of mechanical engineering, gates, valves, piping systems, penstocks, siphons, turbines, pumps, and closed-conduit features are studied. When an investigation is initiated by a design branch, the designers usually submit preliminary designs and specify which features are to be studied for hydraulic performance. In some instances the request is general rather than specific. In either case, the studies, especially when a model is involved, are conducted to obtain the maximum amount of pertinent information.

Model studies are conducted according to mathematical relationships and rules of hydraulic similitude of the model and the corresponding field structure. Comparison of model and prototype has clearly demonstrated that, with few exceptions, there is an excellent correspondence of behavior.[1] Where agreement at first appeared to be lacking, failure to recognize or interpret model results correctly was usually the cause for the disagreement.

Agreement between the model and prototype has been especially complete for overfall spillway crests, valves, and gates. Thus, calibration curves for the operation of these features are generally based on model results in lieu of field calibration. Energy dissipators, including stilling basins and buckets of various types, designed on the basis of model findings, have also been successfully operated in substantial agreement with model predictions.

River improvement plans of large magnitude have worked out successfully according to predictions based on model tests. The high

---

[1] "Conformity Between Model and Prototype in Hydraulic Structures," A Symposium, *Transactions of the American Society of Civil Engineers*, vol. 109, pp. 3-193, 1944.

efficiencies and generally smooth operating characteristics of large modern turbines and pumps can also be attributed to model experiments. In practically all cases where comparisons have been possible, improvements indicated by models have been substantiated by the prototype structures.

CHAPTER 1

Laboratory Equipment

1-1. Laboratory Arrangement. –The Hydraulic Laboratory of the Bureau of Reclamation is housed in a low, sprawling building at the E&R Center (Engineering and Research Center) near Denver, Colorado (fig. 1-1). The laboratory floor space, exclusive of offices, shops, and storage, embraces an area of 4900 square meters, obstructed only by 254-millimeter (10-in) steel columns on 9.14-meter centers. Seventy percent of this area has 7.6 meters of headroom; the remainder, 4 meters. Figure 1-2 shows the arrangement of the laboratory.

Figure 1-1.–The Bureau of Reclamation's Hydraulic Laboratory.
P801-D-79240

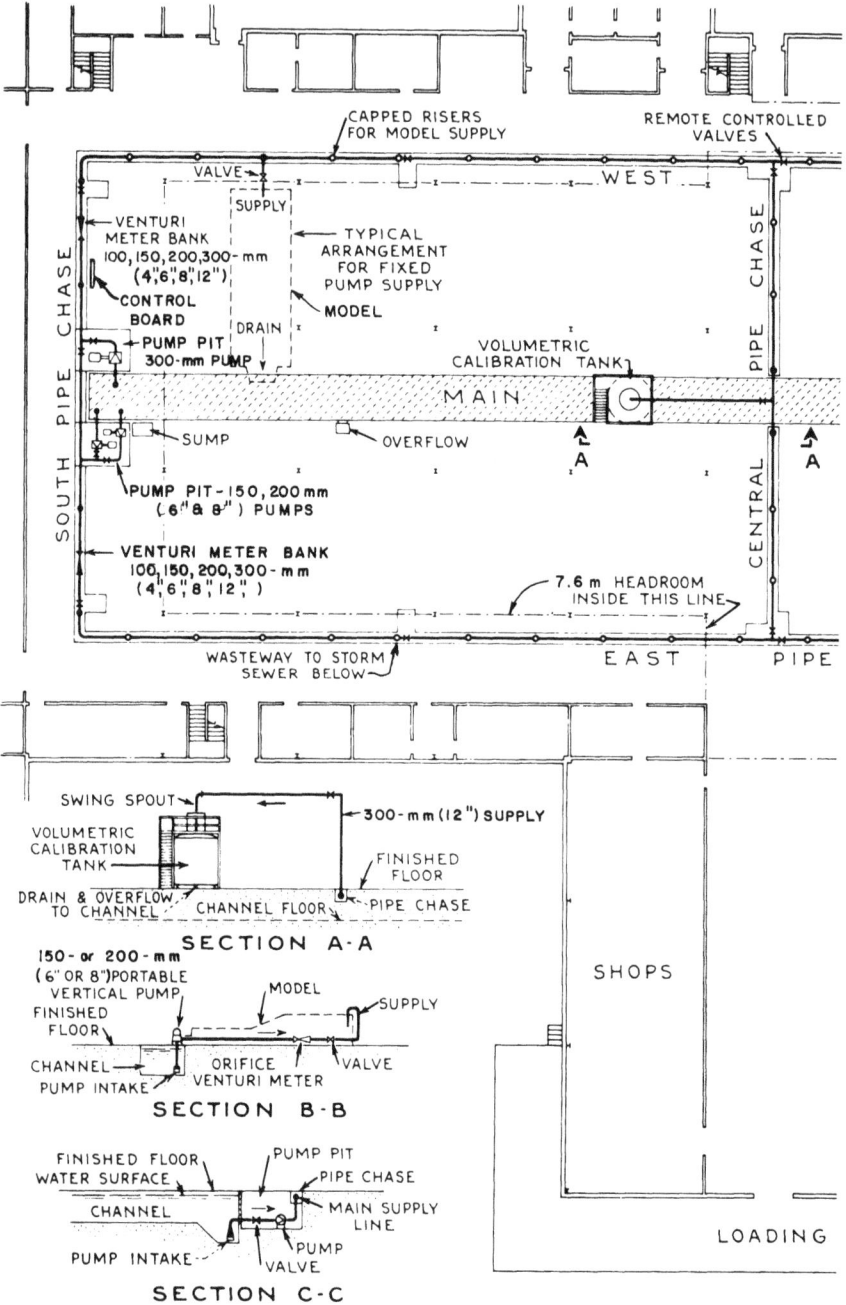

CAPPED RISERS
FOR MODEL SUPPLY

REMOTE CONTROLLED
VALVES

VALVE

SUPPLY

WEST

VENTURI
METER BANK
100,150,200,300- mm
(4",6",8",12")

CONTROL
BOARD

PUMP PIT
300- mm PUMP

DRAIN

TYPICAL
ARRANGEMENT
FOR FIXED
PUMP SUPPLY

MODEL

VOLUMETRIC
CALIBRATION TANK

MAIN

SUMP

OVERFLOW

A

A

PUMP PIT - 150,200 mm
(6" & 8") PUMPS

VENTURI METER BANK
100,150,200,300- mm
(4",6",8",12")

7.6 m HEADROOM
INSIDE THIS LINE

WASTEWAY TO STORM
SEWER BELOW

EAST

PIPE

SOUTH PIPE CHASE

WEST PIPE CHASE

CENTRAL PIPE CHASE

SWING SPOUT

300- mm (12") SUPPLY

VOLUMETRIC
CALIBRATION
TANK

FINISHED
FLOOR

DRAIN & OVERFLOW
TO CHANNEL

CHANNEL FLOOR

PIPE CHASE

SECTION A·A

150- or 200- mm
(6" OR 8")PORTABLE
VERTICAL PUMP

MODEL

SUPPLY

FINISHED
FLOOR

CHANNEL

PUMP INTAKE

ORIFICE
VENTURI METER

VALVE

SECTION B·B

SHOPS

FINISHED FLOOR
WATER SURFACE

PUMP PIT

PIPE CHASE

CHANNEL

MAIN SUPPLY
LINE

PUMP INTAKE

PUMP

VALVE

SECTION C·C

LOADING

2

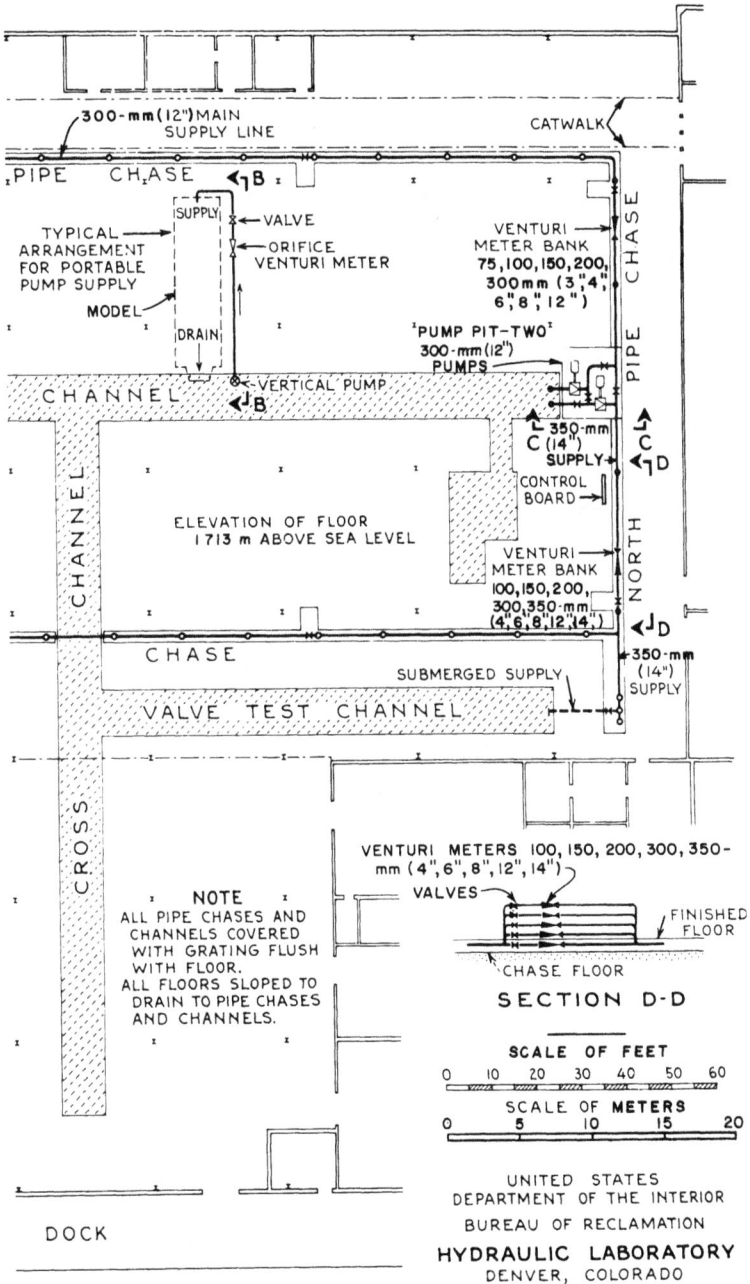

Figure 1-2.–Floor plan of the Hydraulic Laboratory. 103-D-1704.

1-2. **Facilities for Supplying Water.**-(a) *Fixed Equipment.*-Water channels, 2.90 meters wide and ranging from 1 to 2.50 meters deep, are built into the laboratory floor and serve as a reservoir both for the laboratory pumps and for the water discharged from the models. Pump pits located at the ends of the main channel contain three 300-millimeter (12-in) horizontal pumps: two in the north pit, with capacities of 280 and 350 liters per second, driven by 75- and 112-kilowatt electric motors; and one in the south pit, with a capacity of 280 liters per second, driven by a 75-kilowatt motor. The south pit also contains 150- and 200-millimeter (6- and 8-in) pumps. Hydraulically operated gate valves and the piping arrangement in the north pump pit (fig. 1-3) enable those two pumps to be operated in series, in parallel, or separately. Control boards containing pushbuttons, switches, position indicators, gages, mercury manometers, and recorders are located adjacent to each pump pit (fig. 1-4).

Figure 1-3.-North pump pit. These two pumps have individual capacities of 280 and 350 liters per second and may be operated in series, parallel, or separately. P801-D-79239

Figure 1-4.–Control board. P801-D-79238

A pipe chase around the perimeter of the main portion of the laboratory (fig. 1-2) contains the fixed measuring equipment, supply piping, and valves. For drainage, the laboratory floor slopes slightly toward the chases and channels, which are covered with steel grating set flush with the floor. The equipment for measuring pump discharge includes four banks of venturi meters ranging from 75 to 350 millimeters in diameter (figs. 1-2 and 1-5). Piping in the chases is principally 300-millimeter (12-in) standard pipe with tee connections and vertical risers at 4.6-meter intervals. To connect a model to the system, lightweight pipe is installed between the shutoff valve on the vertical riser and the model.

Flow in the fixed lines is regulated with hydraulically controlled gate valves, all located below the floorline except for the venturi meter line valves. The operating cylinders for the main gate valves in the venturi meter lines were especially designed and manufactured for the laboratory water control system. These cylinders are operated by pressure from the building water supply (approximately 480 kPa, gage pressure) and controlled by

5

Figure 1-5.–Venturi meters and hydraulically controlled throttling valves in pump discharge lines. P801-D-79237

pneumatic pilot valves. Opening or closing the main gate valves in the venturi lines, or holding them in any intermediate position, can be accomplished from either of the two control boards. The valve positions are indicated on the control boards as follows: As the stem of a gate valve moves, the linear motion is translated into rotary motion by a screw thread. The screw rotates a potentiometer, and the resulting change in voltage is applied to a meter on the control board which is calibrated in percentage of valve opening.

The remaining gate valves in the main circulation loop have solenoid-operated, four-way valves attached to hydraulic cylinders. The solenoid valves are electrically connected to the control boards and are operated from the boards through microswitches. These gate valves are either open or closed; they have no control for intermediate positions. Individual indicator lights on the control boards show whether the main line valves are open or closed.

Pot-type mercury manometers on the control boards are used to indicate the differential head across the venturi meters. All parts of the manometers that come in contact with the mercury are either stainless steel or glass to prevent their amalgamation with the mercury. The mercury is contained in a stainless steel pot at the bottom rear of the manometer gage. The reading glass is 13-mm-inside-diameter, heavy-wall tubing, which is sufficiently large to minimize capillarity and to increase the damping effect of the connecting tubing. The height of the mercury is read in millimeters from a vernier gage to the left of the glass; gage readings are accurate to 0.06 of a millimeter. Parallax in reading is minimized by leveling the eye with a line on a finder. A small light on the finder illuminates the manometer glass and a portion of the reading scale.

Purge valves are mounted near the top of the manometer. When the valves are open, water flows under steady pressure from the building supply through the gage to the venturi meters by way of two connecting lines. The advantage of this feature is that the meters can be bled of air during operation without having to close the venturi control valve to develop sufficient purging pressure. The gage cannot be bled improperly, and the overflow pot prevents loss of mercury.

One manometer is provided for each venturi meter bank. The pressure taps for each meter are connected to the control board through a pair of 10-mm-diameter stainless steel tubes. A switch on the control board opens solenoid valves in the appropriate connecting tubes to select the meter desired for operation.

(b) *Portable Equipment.*–A versatile piece of equipment in the laboratory is a portable pump unit (fig. 1-6) equipped with a flow-measuring device. This pump can be transported and assembled wherever desired along the water-supply channel. The portable equipment includes a vertical-turbine-type pump, a section of lightweight pipe containing a flow straightener, and a 200-millimeter combination venturi/orifice meter developed especially for this application. The meter (fig. 1-7) has a ring seal that automatically seals the orifice plate in place when the pump is

Figure 1-6.–Vertical portable pump unit and meter (without differential manometer). P801-D-79236

7

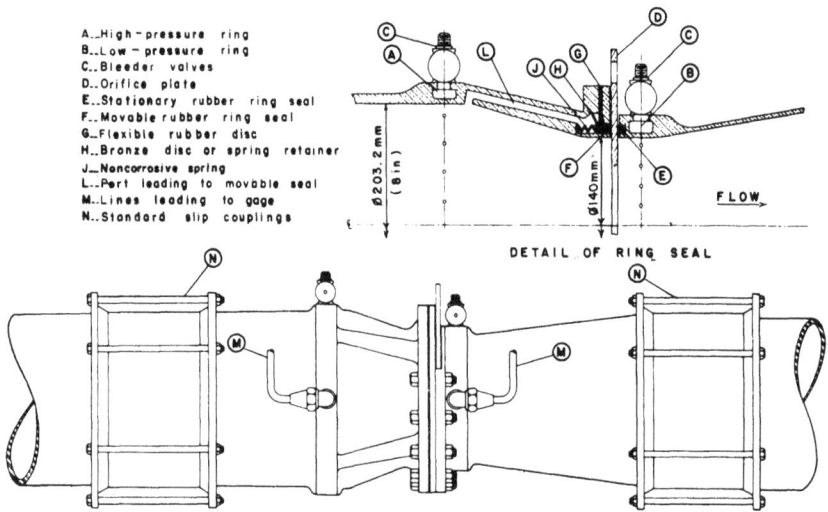

A..High-pressure ring
B..Low-pressure ring
C..Bleeder valves
D..Orifice plate
E..Stationary rubber ring seal
F..Movable rubber ring seal
G..Flexible rubber disc
H..Bronze disc or spring retainer
J..Noncorrosive spring
L..Port leading to movable seal
M..Lines leading to gage
N..Standard slip couplings

Figure 1-7.–Venturi/orifice meter. 103-D-1705.

set in motion. To change orifices or to use the device as a venturi meter, the pump is shut down and the orifice plate is removed from the slot and replaced with the one desired. There are no bolts or clamps to loosen, and it is not necessary to drain the water from the line when changing orifice plates. The device has the advantage of being usable either as a venturi meter for large discharges or as an orifice meter for intermediate and smaller flows; however, the main advantage of this meter is its portability.

Other portable equipment in the laboratory includes 200-millimeter pumps with individual capacities of 140 liters per second and 150-millimeter pumps having capacities of 60 liters per second.

(c) *Calibration Apparatus.*–Venturi meters in the laboratory are calibrated and checked in place by using the volumetric calibration tank located in the center of the laboratory (fig. 1-8). The tank is also used to calibrate other laboratory and commercial meters.

The main tank has a volume of 19 200 liters (678 ft³) and, to save space, it encloses a similar but smaller tank of 2490-liter (88-ft³) capacity (fig. 1-9). These tanks, resembling pipettes having large bodies and small necks, were calibrated using a third pipette tank of measured volume. A satisfactory timing interval can be obtained for discharges of 60 liters per second or less using the small tank; for larger discharges, up to about 340 liters per second, both

8

Figure 1-8.–Volumetric calibration equipment. P801-D-79234

tanks are used. A swing spout diverts the inflowing water across a knife edge at a uniform rate into the small tank, the large tank, or back to the laboratory reservoir.

To calibrate a venturi meter, a steady flow is established with the swing spout in the waste position. The spout is then shifted by a pneumatic jack to one of the tanks until the water reaches a level in the neck that is measurable by hook gage, at which time the spout

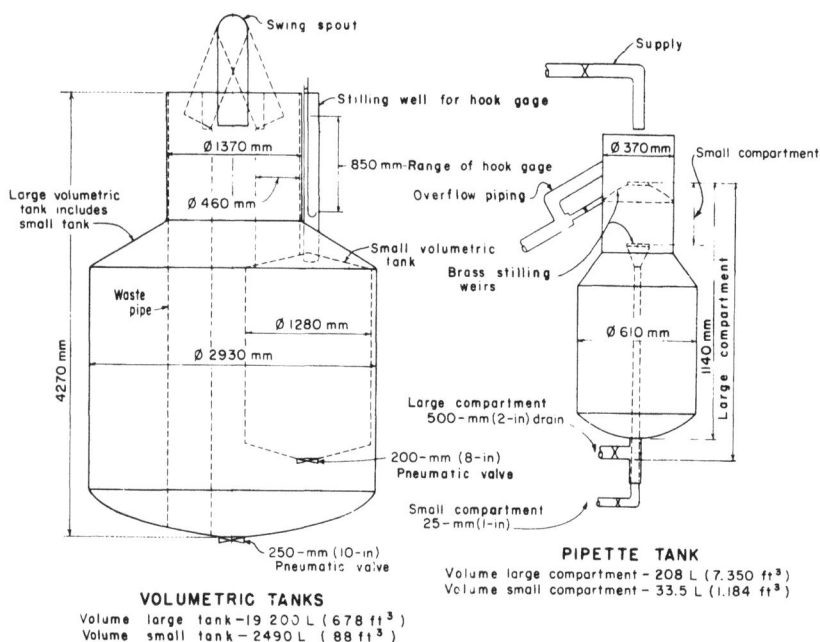

Figure 1-9.–Details of calibration tanks. 103-D-1706.

is returned to the waste position to discharge into the reservoir. During the filling period, time is measured electronically and readings are taken from the mercury manometer for rating the venturi meter. The volume of water in the calibration tank is determined from a graph which shows the volume of the tank for any hook gage reading and temperature. The pneumatic valves on the tanks are interlocked to prevent filling the large tank with the valve in the small tank closed. If this interlock was not provided, there would be a tendency to float the small tank, resulting in distortion of both tanks.

1-3. Special Testing Facilities–(a) *High-Head Test Facility.*–To provide the capability of performing laboratory tests at actual prototype heads, a pump which will develop heads in the range of 75 to 185 meters of water was installed in the laboratory (fig. 1-10). This 200-millimeter, seven-stage, vertical-turbine pump is driven by a 185-kilowatt, direct-current motor. Auxiliary equipment includes: a rectifying unit, a motor speed control which receives feedback signals from a tachometer mounted on top of the

10

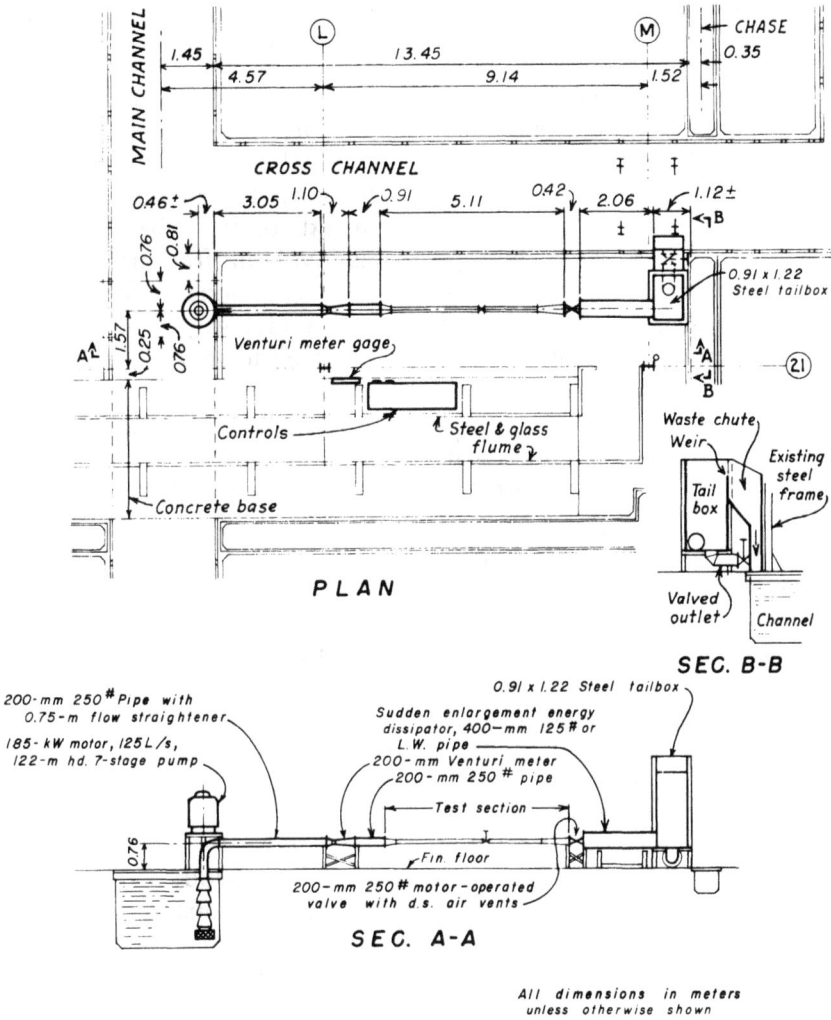

Figure 1-10.–High-head test facility. 103-D-1707.

motor, and a manually operated rheostat for speed selection which is infinitely variable from 200 to 1800 revolutions per minute.

Rate of flow is measured with a 200- by 110-millimeter venturi meter permanently installed 3 meters (15 diameters) downstream from the pump outlet. A 750-mm-long flow straightener is provided in the pipe at the pump outlet to reduce spiraling of the flow as it passes through the meter. The meter was calibrated in place using the volumetric calibration tank described in the preceding paragraphs.

11

Performance characteristic curves for the system (fig. 1-11) were developed from the laboratory acceptance tests.

(b) *Glass-Paneled Flume.*–A steel-framed, glass-walled flume (fig. 1-12) is a fixed part of the laboratory test equipment. This flume may be used for studying sectional models of spillways, aprons, and energy dissipators. The glass panels allow visual inspection of the action of suspended loads or bedloads in sedimentation studies. Many problems involving free-surface flow may be studied, including those related to erosional tendencies at hydraulic structures and to flow over or under gates and spillway crests. The flume may also be used as a tank to study wave forces on embankments, dikes, or riprap slopes.

The flume is 24 meters long and is constructed of panels 3 meters long and 2.4 meters high anchored to a concrete floor. Panels are interchangeable: either nine tempered-glass windows, 13 millimeters thick, or a single steel plate may be installed in any panel. A reinforced concrete slab containing access pipes for piezometer lines and holes for anchor bolts serves as a floor for the flume. This slab rests on the laboratory floor and is sufficiently wide to provide support for steel buttresses located at the ends of each panel.

The head box, used to quiet the inflow of water to the flume, is 3 meters high–0.6 meter higher than the flume to allow for surges and head loss through the flow-straightening baffle. The baffle consists of sheets of 380-mm-wide corrugated steel, placed horizontally and 25 millimeters apart, with the corrugations perpendicular to the flow. Water passing through the baffle undergoes a series of expansions and contractions between the corrugations and enters the test flume uniformly with very little head loss. Flows ranging up to 930 liters per second can be pumped through the flume.

Elevation of the water surface in the flume can be regulated by various types of controls at the downstream end of the flume. These controls are built as needed and depend on the type of model test to be conducted.

(c) *Electrical Analogs.*–Conducting-paper and water-tray electrical analogs are used for preliminary laboratory studies. Examples of subjects that may be studied by the analog model are: the shaping of flow passages, the need for flow straighteners, relative seepage losses from canals, flow coefficients for draintile, and cavitation tendencies along a flow boundary. In a two-dimensional flow problem, a satisfactory answer may be

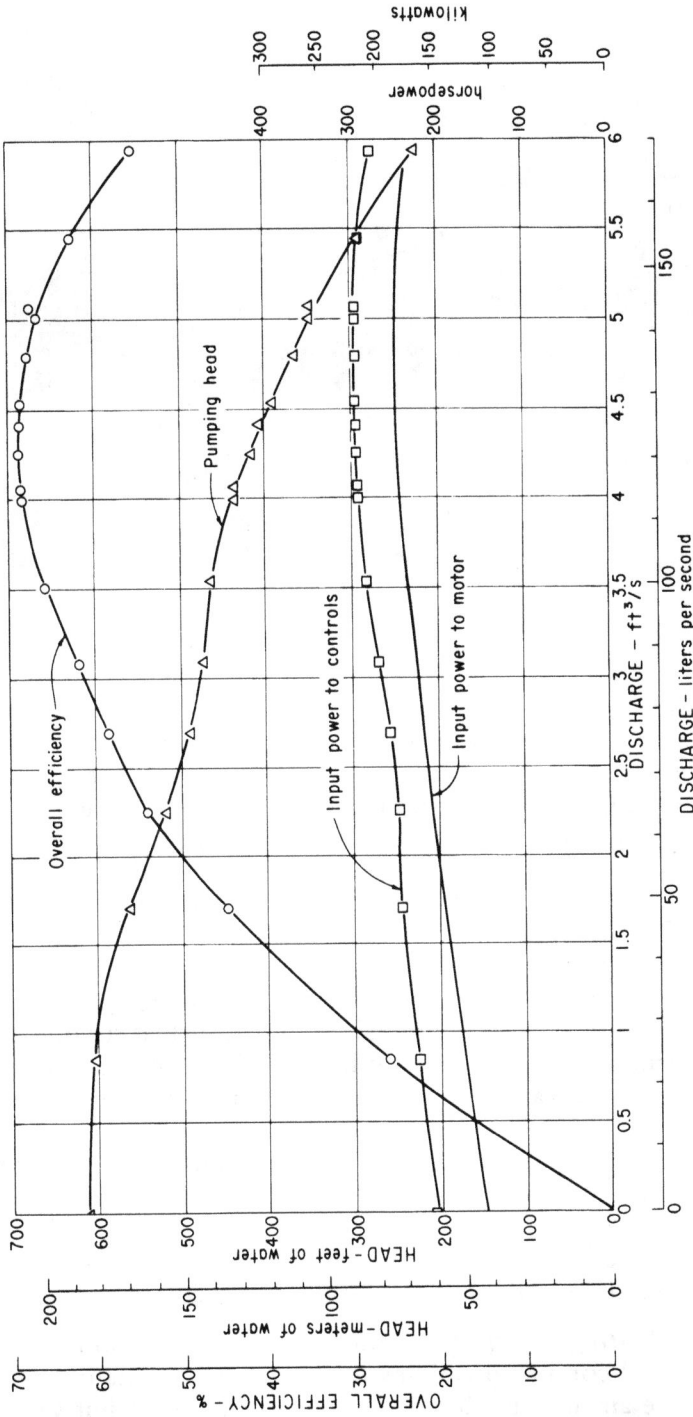

Figure 1-11.—High-head pump performance curves. 103-D-1708.

13

Figure 1-12.–Glass-walled test flume (flow left to right). P801-D-79235

obtained from the analog alone. For studies of flow in three dimensions, the analog is used to supplement the hydraulic model.

A satisfactory electrical analog system may be comprised of carbon-coated paper, silver paint, a precision 10-turn potentiometer with a scale reading to 100 parts per turn, a null-type milliammeter, and a low-voltage, direct-current power source. For studies of large areas (greater than the available 508-mm-wide paper will provide), a water tray and metal electrodes may be substituted for the carbon-coated paper.

Modern analog computers are available for solving problems involving hydraulic transients. Special problems, such as surging in low-head irrigation pipelines, may be partially solved using an analog with resistive/capacitive elements simulating the distribution system.

Ground-water, tidal, watershed, and flood-routing studies have been performed with sophisticated electrical analogs. It may be possible to use digital mathematical models in studies involving large, complicated hydraulic systems; however, electrical analogs should be considered first in laboratory investigations.

(d) *Air-Testing Equipment.*–Air as a model test fluid is used to solve closed-conduit problems involving valves, gates, and pipelines. The air model is better adapted to closed-conduit work

14

but can also be used in free-surface studies by selecting suitable points for measuring the effective head. Air testing has the advantages of low cost, dryness, speed, and ease of model modification. In air testing, it is possible to construct models of plaster of paris or wood and to complete the testing before similar hydraulic models could be constructed of more durable materials. A centrifugal blower, with a capacity of approximately 1.4 cubic meters per second at 230 millimeters of water, supplies the air. The blower (fig. 1-13) is driven at a constant speed, and the total airflow is measured by sharp-edged intake or discharge orifices. A discussion on model testing with air is in Chapter 8, Closed-Conduit Flow.

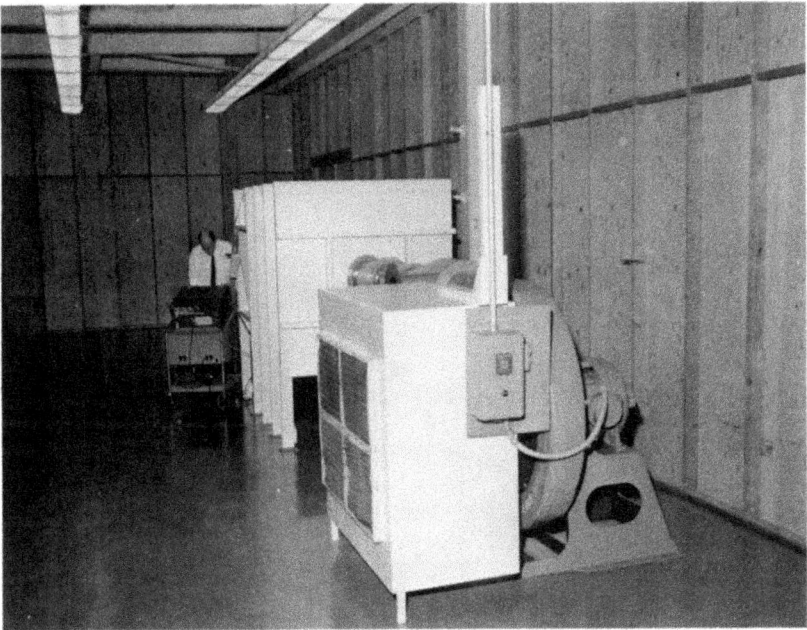

Figure 1-13.–Centrifugal blower for studies using air as the test fluid. P801-D-79232

(e) *Sediment-Feeding Apparatus.*–A special device for providing a uniform supply of sediment to movable-bed models is shown in figure 1-14. A sheet metal hopper having the shape of a prism is mounted so the opening in the bottom is just above a pan on a vibratory-type feeder. The feeder is a trough or pan mounted on flexible leaf springs supported in a frame and vibrated at 60 hertz by an electromagnet. The magnet pulls the trough sharply down and back; then, when the current is reversed, the leaf springs return the

15

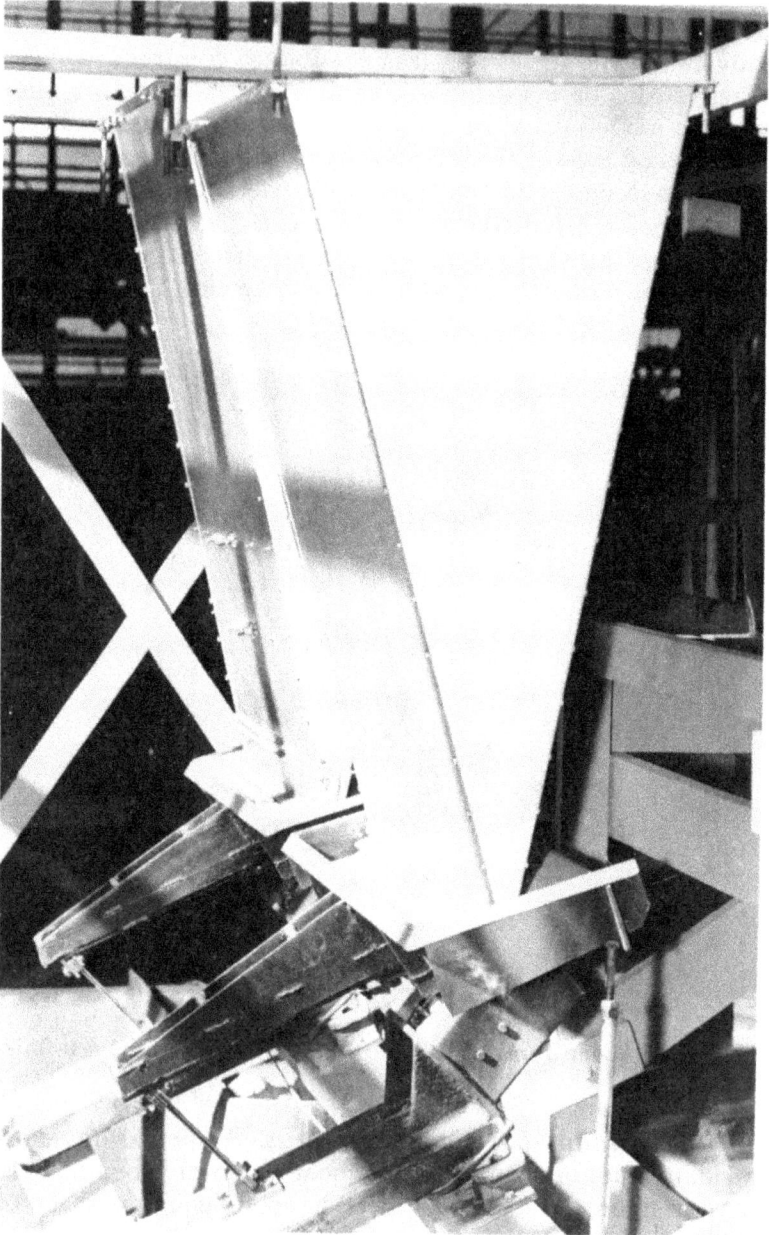

Figure 1-14.–Apparatus for uniform feeding of sediment to movable-bed models. P801-D-79233

pan up and forward to the starting position. Because the return is not as sharp as the downward pull, the resulting motion causes material on the pan to move forward even when the pan is sloped upward.

The feeder is equipped with a separate electrical control box containing the operating switch, a rheostat for controlling the vibration amplitude and thus the rate of flow of sediment, and a circuit to generate sharp pulses from the alternating current. The pulses activate the magnet and cause the vibratory motion of the feeder. Sediment that has a mean size smaller than 0.50 millimeter must be dried before it is put in the feeder; larger sediment will feed satisfactorily when damp.

(f) *Constant-Head Tank.*–The nature of the majority of investigations performed in the Bureau of Reclamation Hydraulic Laboratory makes the use of a constant-head tank unnecessary. Large head boxes, with gravel baffles to calm the water, absorb the relatively small discharge variations produced by the centrifugal pumps and piping. When constant heads and extremely steady flows are necessary, special tanks and boxes are provided for the study.

(g) *Low-Ambient-Pressure Chamber (Vacuum Tank).*–A recently constructed laboratory facility allows testing of models of open channels and closed conduits under scaled-down atmospheric pressure. Research into formation of cavitation and protection against cavitation damage is conducted in the facility, as well as studies in which it is desirable to cause cavitation to form in a model in order to predict its occurrence in the prototype.

CHAPTER 2

Instrumentation

2-1. **General Considerations.**–Since the object of every model investigation is a carefully planned series of measurements, the instruments required to make these measurements comprise an essential feature of the laboratory equipment. Laboratory studies involve measurement of static and dynamic quantities. Simple instruments (point gages, pitot tubes, and manometers) are used for static or slowly changing hydraulic measurements. For dynamic changes, electronic equipment is required to produce a satisfactory measurement. Considerations involved in using instruments of both static and dynamic types are described in this chapter.

2-2. **Discharge Measurements.**–Laboratory devices for measuring discharge include standard weirs, venturi meters, orifices, rotameters, disk-type water meters, and flow nozzles. Volumetric and weighing tanks may be included, although these usually serve as calibration equipment for checking the accuracy of standard meters. Weirs can be constructed by following the recommendations contained in any good hydraulics textbook. Venturi meters, being more difficult to build, are usually purchased from a meter company. Orifice meters and flow nozzles can be purchased from commercial sources, or they can be constructed in the laboratory by adhering to the specifications set forth by the American Society of Mechanical Engineers [1].[1] Intake orifices [2] and discharge orifices [3] are also useful measuring devices in a laboratory. Other discharge-measuring devices used successfully in a laboratory are the elbow meter [4], the Parshall flume, the propeller-type flowmeter, and the current meter [5]. The devices

---

[1] Numbers in brackets refer to entries in the bibliography.

should be calibrated in place using suitable weighing or volumetric tanks. When in-place calibration is not possible, commercial meters or laboratory-constructed meters of standardized designs may be used. Flowmetering devices are chosen for laboratory use according to the desired accuracy of the measurement.

2–3. **Measurement of Water Surface Elevation.**–The elevation of a water surface is usually measured with a point gage over the surface or with a hook gage in a stilling well which is proportioned to dampen depth fluctuations transmitted to the well. Point and hook gages (fig. 2-1) are operated through rack-and-pinion mechanisms and are available commercially in various lengths. These gages may be used to indicate differences in elevation or absolute elevation. By use of special mountings, they are adaptable to a variety of conditions; if mounted on rails, a large area can be covered. Float gages connected to an indicating mechanism consisting of a pulley, large dial, and needle will measure small changes in water surface elevation.

2–4. **Velocity Measurements.**–Accurate measurement of velocities in laboratory investigations requires good instrumentation. Instruments appropriate for moderate or high velocities may not be suited for low velocities, and those designed for low velocities may be too fragile to withstand high velocities. Some instruments are more suitable for measurement in open channels than in closed conduits.

The most common velocity-measuring devices are pitot tubes, some of which are shown in figure 2-1. They may be readily fabricated from hypodermic needle tubing and are made in various sizes, depending on the work requirements. Pitot tubes B and D are modified Prandtl types. Tube B has both kinetic and static legs and is designed for a coefficient of unity. This type is widely used because it works well over a large range of velocities. Because of its shape, it is principally applicable to open-channel measurements. Tube D has a total-head opening only and is used for measuring air and water velocities.

Tube C is a pitot cylinder suited for measuring medium and low velocities in closed conduits. It has three ports in a plane normal to the centerline of the shaft. The center or total-head port is pointed directly into the flow by rotating the pitot tube until pressures on the two outer ports are balanced. For its design range, the tube registers approximately 1.33 times greater than the theoretical velocity head. Tube C is supported at both sides of the conduit through packing glands to allow velocity traversing across the conduit. Velocity distribution studies are possible when packing

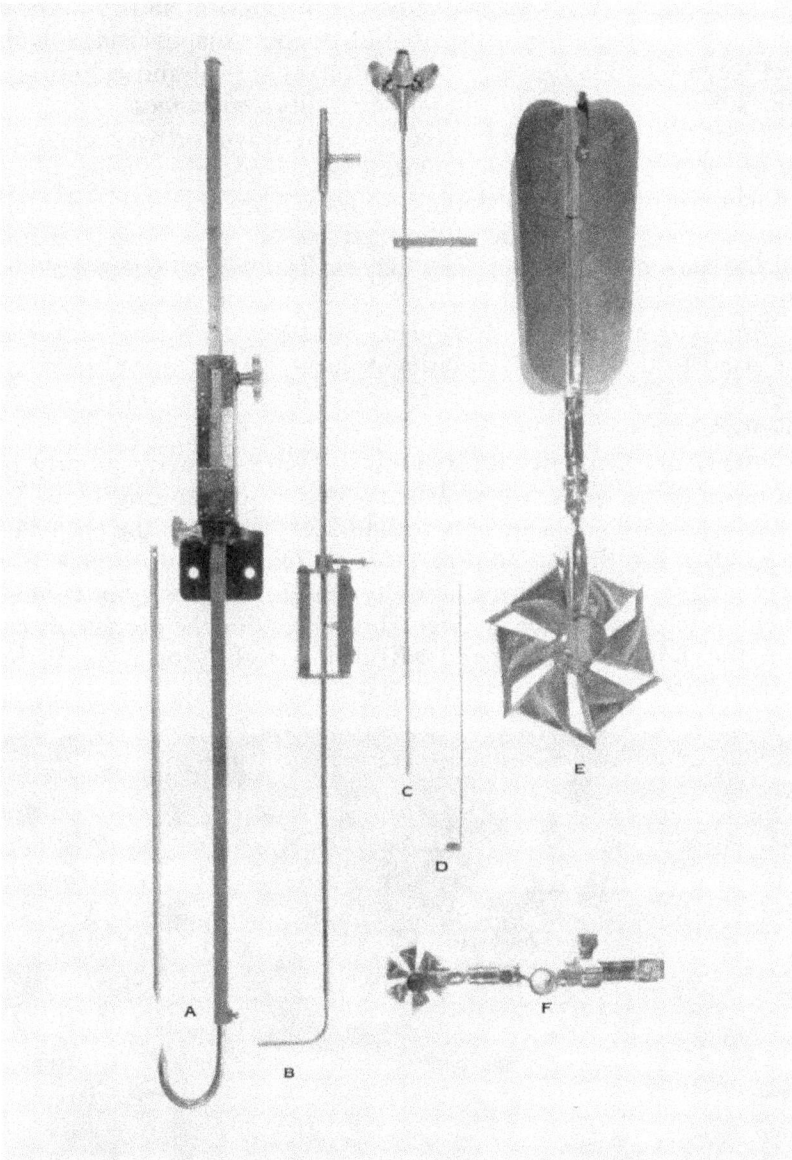

Figure 2-1.–Instruments for measuring water surface elevations and velocities. A–point and hook gage; B, C, and D–Prandtl, cylindrical, and total-head pitot tubes; E and F–current meters. P801-D-79231

21

glands are provided at opposite ends of several diameters in the same cross section of conduit.

Coefficients vary with the velocity for spherical and cylindrical pitot tubes, as the pressure distribution on these shapes changes with the Reynolds number. For this reason, spherical and cylindrical tubes must be used only within their calibrated ranges.

Pitot tubes may be mounted in many ways for open-channel work. The support may be rigid or it may be capable of being tilted to measure velocities on sloping surfaces. A micrometer screw can indicate the vertical angle of the shaft and allow for some rotation of the pitot tube. The supports may be designed to operate along structural channels.

Price current meters of standard or pygmy sizes (fig. 2-1) may be used to determine velocities in large open-channel models. Screw-type current meters are also used in laboratory measurements.

Miniature current meters (fig. 2-2) are used frequently in open-channel work for measuring velocities ranging from 30 millimeters to 1.50 meters per second. Revolutions of the propeller are electronically counted for a period of time and displayed or recorded on a digital tape printer to indicate the flow velocity. The meter is sensitive to flow in either direction along the propeller axis, but the velocities measured will differ slightly because of the bearing arrangement.

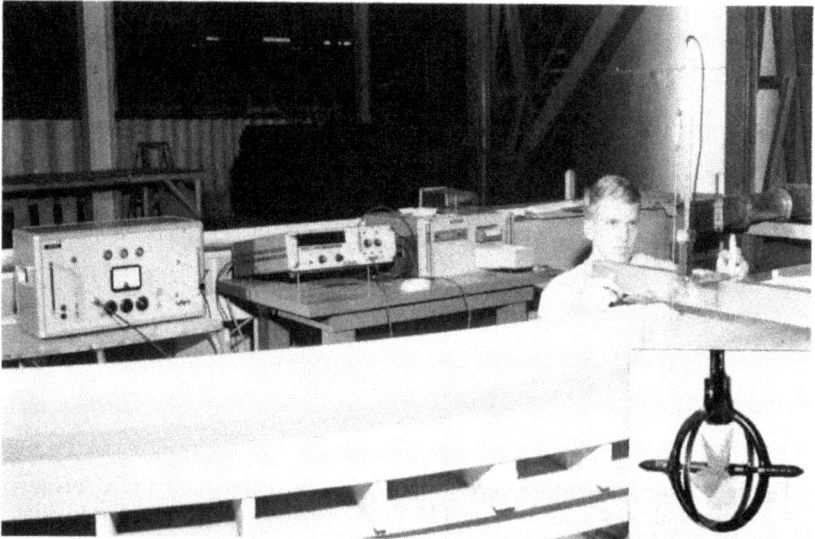

Figure 2-2.–Velocity meters. Current meter with a 10-mm-diameter propeller. (Pulse counter and printed tape recorder in background.) P801-D-79247

Another device used for measuring water velocity in the laboratory or the field is a current meter based on the Faraday principle of electromagnetic induction. The electromagnetic current meter includes a flow sensor (an electromagnetic coil and electrodes) and signal converter with analog output. The voltage induced by the flowing water is directly proportional to the velocity. Either one or two pairs of electrodes may be contained in the wall of the sensor. Two pairs of electrodes provide analog voltages linearly proportional to the X and Y components of the velocity vector. The sensing head used in the Bureau's laboratory has been miniaturized, permitting velocities to be measured throughout the flow area. The meter averages the velocity in a flow field of about three diameters of the 13-millimeter (1/2-in) sensing cylinder or sphere. Current meters are supplied with three switchable ranges: plus or minus 60, 150, and 300 centimeters per second. The smallest velocity the user could expect to measure is approximately 15 millimeters per second.

Hot-wire and hot-film anemometers are used to measure velocity and turbulence in studies utilizing either air or water as the working fluid. A wire or a film sensor is heated to a controlled temperature while immersed in the fluid. An increase or decrease in the flow velocity past the sensor causes the rate of heat conductance to change accordingly. The change in electric current required to maintain the controlled temperature can be related to the variation in velocity.

Other anemometers are available commercially or they may be fabricated in the laboratory. Thermistors, devices that change resistance with temperature, may be adapted to velocity measurement. Small pressure transducers may be combined with pitot tubes to study turbulent fluctuations of velocity. Averaging circuits in the electronic instruments used with the transducers can also be used to obtain the mean and root-mean-square velocities. The researcher must continually devise new methods for measuring velocity in model and prototype studies.

Additional methods have been developed for indicating velocities. In one of these, particles of powdered aluminum, paper confetti, hydrogen bubbles, dye traces, or tellurium oxide are introduced into the fluid and photographed. A length scale must be placed on the model so that it also appears in the photograph. The lengths of the particle traces are then scaled from the photograph, and because the shutter speed of the camera is known, velocities at various points in the model can be computed.

A modern technique of velocity measurement involves the use of a laser and optics (lenses and mirrors) to measure point velocities in a fluid. The difference in frequency between a reference beam and laser light reflected from particles in the flow constitute the Doppler shift, which can be converted to velocity. This method has the major advantage that all instrumentation is outside the flow, which remains undisturbed. The major disadvantage at this time (1980) is cost.

2-5. **Pressure Measurements.**–The simplest device for measuring positive water pressures (above atmospheric is the single-leg manometer shown in figure 2-3. Each glass ube is connected by flexible tubing to a piezometer tap located on the model at the point where a pressure measurement is desired. The pressure is then read directly in millimeters or meters of water above that point. When using small tubes, a drop of wetting agent (similar to that used in photograph surface finishing, or detergent) should be inserted in the tube to reduce the surface tension. To improve definition of the meniscus, the water in the tubes is often colored by adding a trace of fluorescein or other nonstaining dye.

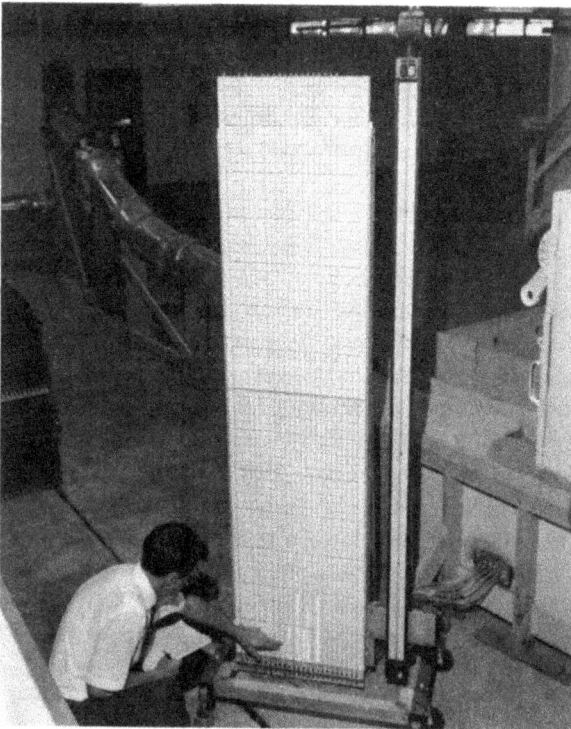

Figure 2-3.–Measuring pressure with manometers.
P801-D-74594

24

The U-tube manometer can be used to measure negative as well as positive pressures. Various liquids, including water, mercury, alcohol, and specially prepared commercial compounds, may be used in this manometer; the choice depends upon the magnitude of pressures to be measured.

Improperly constructed piezometers give false readings; therefore, a single piezometer should not be relied upon for accurate pressure measurement. Piezometers should be installed in groups, and pressures should be measured with manometers that may all be viewed simultaneously (fig. 2-3); a pressure that appears to be in error may then be checked immediately on the manometer board. The practice of using one manometer and connecting it consecutively to different piezometer taps should be avoided because faulty readings are difficult to detect when this technique is employed.

Average values of measured pressures may be obtained from repeated manometer readings. Manometers connected to piezometers of the same group should be read simultaneously to ensure the proper interpretation of the relationship between pressures in adjacent areas. Simultaneous readings may be obtained by pinching off all connecting tubes at the same instant with a mechanical device or by photographing the manometer board. The photographic method is preferred because the manometer readings may be checked, if necessary, at a later date.

The gage on the right in figure 2-3 is a pot-type manometer in which a reservoir is substituted for one leg of a U-tube manometer. It is designed to read negative, positive, or differential water pressures directly on one scale. For large positive water pressures, mercury may be used in the manometer; however, for the small positive water pressures encountered in the laboratory, single-leg water tubes are preferred.

The gages in figure 2-4A are well-type manometers used for measuring positive, negative, or differential water heads. Movable wells enable measurement over a greater range than would be possible with fixed reservoirs. Bubbles on the frames may be used to level the gage for accurate measurements. Micrometer screws and point gages allow estimating pressure head to 0.025 millimeter. A trace of wetting agent is added to the water in the wells to decrease surface tension.

For measuring positive pressures of up to 30 meters of water, mercury U-tubes or pot gages are usually employed. For higher pressures, a fluid-pressure scale (fig. 2-4B) is available. The scale is an adaptation from a deadweight testing machine in which the force of fluid pressure against a piston, which actuates a system of

A. Null-type pressure or vacuum manometer (left) and differential point gage. P801-D-74589

B. Fluid pressure scale for "weighing" pressure up to 300 pounds per square inch (2000 kPa). P245-D-60432

Figure 2-4.–Laboratory instruments for measuring pressure.

levers, is measured as if it were a mass. By loading the main lever arm with weights calibrated in pounds per square inch until a balance is obtained, the pressure can be determined. A motor revolves the piston slowly, thus minimizing static friction between the piston and the walls of the cylinder. Pressures up to 300-pounds-per-square-inch gage (2000 kPa) can be measured with the instrument illustrated and up to 4000-pounds-per-square-inch gage (27 500 kPa) with another instrument similar in design.

Bourdon-tube gages can often be used satisfactorily for pressure measurements in the laboratory. The quality of the gage determines its suitability for accurate measurements in model or prototype studies [6]. Pressure elements in these gages are usually made from noncorrosive materials, including bronze and quartz. Gages of this type range in quality from approximate indicators of pressure to those having the high accuracy of manometers and deadweight testers. Each gage should be judged on factors such as construction of the pressure mechanism, temperature compensation, ease of maintaining calibration, and the accuracy necessary in the pressure measurement.

2–6.   **Time Measurements.**–Accurate time measurement is particularly important in calibrating laboratory equipment or in rating water measurement devices. Time intervals for model studies of dynamic pressures, vibration, or movement must also be measured accurately. A stopwatch, accurate to about one-fifth of a second, may be used for general timing, but where greater accuracy is required, an electronic counter having a controlled-frequency oscillator is used, either for measuring real time or for precise timing of intervals. The resulting information may be recorded on analog charts in digital or impulse form.

2–7.   **Dynamic Measurements.**–Study of unsteady flow in laboratory or prototype structures requires instrumentation for recording and processing data in short time periods [7]. Selection of instruments is controlled by the type and range of the variable to be measured, by the quantity of data required, and by the method of data analysis. In both hydraulic model investigations and field studies, measurements may include velocities, pressures, water depth, wave characteristics, strain, and vibration amplitudes. Field measurements may, in addition, include temperatures, salinity, oxygen content, sedimentation, ground-water levels, cavitation pressures, and accelerations, and may require special adaptation of instruments.

2-8. Selection of Instruments. –Instrumentation should be considered early in the design of model or prototype studies. Procedures usually include choosing basic components such as sensors, amplifiers, and recorders that are best suited to the investigation. Problems of selection can generally be solved by adapting commercially available instruments. Instrument systems should be capable of being calibrated to 1 percent or better within the selected working range.

Study of new instruments and data acquisition methods should be continual to assure that modern techniques are used for gathering and analyzing hydraulic data. Computer-oriented methods of analysis permit rapid acquisition of large quantities of data; therefore, instrumentation should record or interface information that is usable in modern electronic digital or general-purpose analog computers.

In selecting instruments for hydraulic measurements, their interchangeability should be considered. When available, mountings for the equipment should be chosen that will permit the assembly of systems in instrument racks. A system for a particular study may thus be assembled from individual instruments, and upon completion of that study, the components may be reassembled to form systems for other studies.

Components are chosen to be suitable for a variety of signal levels and not for a particular sensor; thus, they may be used for conditioning the signal from resistive, inductive, capacitive, reluctive, piezoelectric, piezoresistive, or other types of sensors. Selection of the signal-conditioning components should include consideration of future application to computer programing and data analysis. Some of the equipment (fig. 2-5) is specialized, but components of the specialized equipment can be used in other systems.

Recorders are selected according to the information required from the study; both analog and digital types (fig. 2-5A) are available in direct-writing, paper-tape, and magnetic-tape models. The recording equipment should have the capability of converting from analog to digital information and vice versa.

(a) *Transducers.*–The transducer or sensor is the part of the equipment that converts the physical quantity being measured into a signal suitable for recording (e.g., from a pressure head to an electrical signal), and the type selected or developed is of utmost importance in instrumentation since a weak, poorly designed sensor will give questionable results. Selection of the sensor is based on a consideration of range, frequency response, sensitivity,

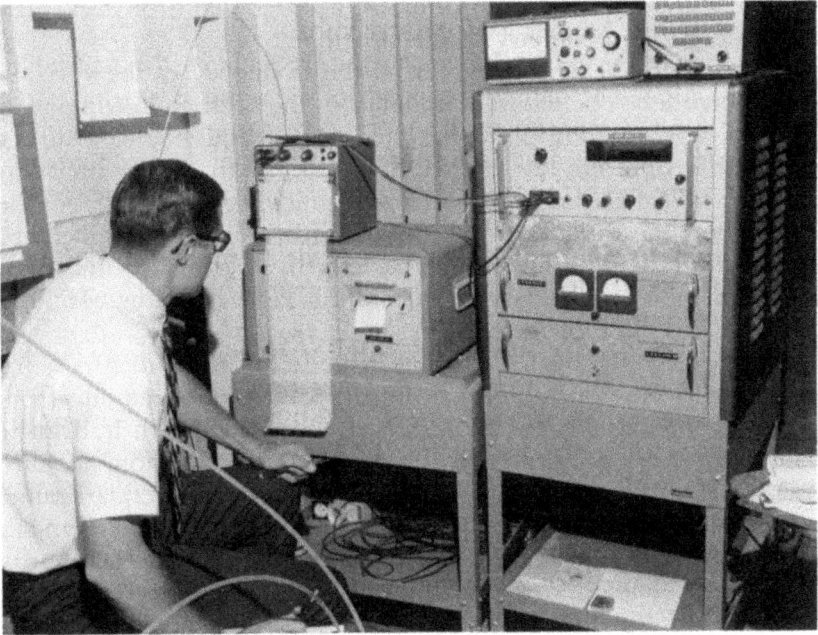

A. Analog and digital measurement of conductivity fluctuations in saline water solution in pipeline. P801-D-79230

B. Measuring pressure fluctuations near entrance to penstock model. PX-D-58747

Figure 2-5.–Systems for dynamic measurements.

temperature stability, environment, accuracy, and physical size required by the geometry of the structure.

Close attention must be given to the frequency of the variable to be recorded. The natural frequency of the sensor's moving parts must be high compared to the measured frequency to assure a negligible time lag and accurate amplification, or attenuation, between signal and recorder. So that the moving parts of the sensor may follow the variations of the applied force, inertia must be kept to a minimum. The materials and lengths of pressure lines and electrical cables must be chosen carefully since they can seriously alter the response of transducers.

Proper sensitivity of the transducer is essential: pressure transducers may be required to measure pressures ranging from units of kilopascals to thousands of kilopascals. In hydraulic laboratory work performed by the Bureau of Reclamation, pressure transducers in the 7- to 175-kilopascal (gage or differential) range are adequate for most studies. It is anticipated that, in time, a collection of transducers will be acquired by the Bureau's laboratory that can be adapted for other studies. Instruments designed for the laboratory frequently are used in the field to make measurements on the prototypes. Temperature compensation, negligible response to acceleration, high resolution, low nonlinearity and hysteresis, and small zero shift are desirable characteristics to consider when selecting a sensor [8, 9].

(b) *Recording Instruments.*–The choice of recording instrument is usually determined by the frequency of the measured variable. Recordings may be in the form of analog or digital information. Water-stage recorders utilizing ink and pen are used to record long-term, slowly changing variations; frequencies up to 100 hertz may be easily recorded using the sensitive recording equipment presently available. Frequencies up to 400 hertz are readily recorded on electrostatic oscillographs. Frequencies higher than 400 hertz may be recorded by light galvanometers, by photographing the trace on the screen of a cathode-ray oscilloscope, or on magnetic tape. Punched- or printed-paper or magnetic-tape recorders can be used to provide information in digital form. A recorder is a precision instrument and should be treated carefully to ensure proper operation.

(c) *Signal Conditioners.*–Signal-conditioning instruments connecting the sensor to the recorder may be mechanical, electrical, or electronic. These instruments supply operating voltages and match the output of the sensor to the input of the recorder. The chief requirement placed on this component is that the signal from the transducer be transferred to the recorder with minimum distortion. Mechanical linkages may be used in recording pressure and temperature. For studies where electrical or electronic linkages are used, the recorder must have a current-sensitive coil, electrodes, or magnetic heads to react to the input signal. Impedance matching is of paramount importance in the selection or design of this linkage to prevent false indications, distortion of the records, and overloading of components. Adjustments must be provided in the linkages to permit calibration and zeroing of the equipment.

Certain types of sensors containing internal signal-conditioning electronics may be connected directly to the recording system. Other types of sensors, such as the geophone used in geological investigations, generate their own voltage. For these sensors, external amplifiers match the impedance of the geophone and provide adequate amplification to operate the recorder. Dynamics studied and recorded in hydraulic testing are generally in the frequency range of 0 to 15 000 hertz. Alternating-current (carrier) or direct-current amplifiers may be used to condition the transducer output before the signal enters the recorder.

The upper limit of dynamic response of a carrier amplifier is generally considered to be about 10 percent of the carrier frequency. Limiting frequency of the system is about 10 percent of the natural frequency of the sensor, or a lower frequency determined by the response of the sensor and connecting pressure lines. Carrier frequencies used in hydraulic testing range from 3000 hertz upward. The natural frequency of the moving part of a transducer is almost always higher than the frequency of the pressure variations measured in hydraulic studies. Studies of cavitation and turbulence require instrumentation having a frequency response ranging to 15 000 hertz and higher.

One of the primary problems in instrumentation results from the effect that the transfer medium has on an electrical signal. In pressure measurements for instance, the amplitude and frequency of the electrical signal may be distorted radically from the applied force by the tubes used to connect the transducer to the source of the pressure change. Short, rigid tubes will normally provide the response necessary for good dynamic measurements. Studies of the frequency and amplitude response of the system should be made to assure the validity of the measurements.

2-9. **Special Instrumentation.**–Although a wide range of instruments is available commercially, occasionally special mechanical and electronic instruments must be constructed to meet the requirements of a particular study. The transducers for measuring waves shown in figure 2-6 are examples of custom-made and assembled equipment.

Analog and digital computers should be considered a part of the instrumentation for laboratory investigations. Instruments assembled for measurement should have an output compatible with the computers available for acquisition and analysis of the data.

An analog model using the electrical equivalent of hydraulic circuits may have a useful place in laboratory investigations. The model may be used to determine the division of flow among the various channels of a delta, the seepage from a canal or under a dam, or the shape of a flow boundary. Corresponding electrical and hydraulic quantities are shown in the following tabulation:

Hydraulic	Electrical
Elevation of water surface (head)	Voltage
Discharge	Current
Oscillation in pressure or water surface	Alternating voltage
Inertia due to mass	Inductance
Frictional resistance	Electrical resistance
Storage of channel	Capacitance

Figure 2-6.–Wave measurement with capacitance-wire transducers. PX-D-63963

The ratios interrelating the analogous hydraulic and electrical quantities should be chosen to permit easy transference; for example, 1 milliampere of current may be selected to correspond to a discharge of 10 cubic meters per second. Special features of an analog have to be designed to properly represent the analogous change in the electrical and hydraulic models.

2-10. **Operation of Instruments.**—After the instruments have been selected or constructed, calibrated, and tested for accuracy and stability, preliminary measurements are made on the model. Study of these records may suggest ways of improving the instrumentation. After refinements are made, the instruments or system are ready for use. Although most instruments are relatively simple, initial operation is assigned to an instrumentation specialist to assure accurate records. Operators of the equipment should have sufficient experience to detect errors that may appear during the study.

Other items of special equipment or instrumentation for use in laboratory and field studies will be discussed in regard to their specific applications.

2-11. **Photographic Equipment.**—Cameras for taking both still pictures and movies are essential in hydraulic studies. Other equipment should include such accessories as wide-angle and telephoto lenses, filters, flashguns, and tripods. High-speed cameras can be useful in revealing flow changes in laboratory and field studies. Photographs of hydraulic actions, conditions of structures, arrangements of instruments, and data are indispensable records.

2-12. **Television Equipment.**—Portable television monitoring, recording, and playback systems can be used advantageously in acquiring information from model or prototype studies. Having flow conditions recorded on video tape allows the investigator to quickly review and compare a current condition with those observed previously. Video tape may also be used for communication with those located away from the study site.

Underwater observations of structure behavior, sediment movement, or flow conditions may be necessary in some model and prototype investigations. Submersible television cameras and remote monitors may be used to make this type of observation, with video tape or photographs of the monitor screen providing a record of the data. Locations not accessible when water is flowing in a prototype structure may be viewed by placing a camera in position before starting the flow.

2-13. Dynamic Measurement Analysis.—Methods of analyzing the acquired measurements should be considered concurrently with the planning of the instrumentation. The arithmetic average of a series of pressure measurements may be satisfactory for tranquil flows. On the other hand, an investigation of a stilling basin may require measurement of fluctuating pressures, vibration, and waves for design information. Special instruments and methods of analysis may be required for correlating measurements of fluctuations from various parts of a structure, thus requiring multiple-channel recording. Data on magnetic tape may be correlated at some scaled time either faster or slower than the acquisition time. The analysis of these measurements may include root mean square, maximum and minimum, and frequency and power spectra.

Analog and digital methods are available in software and hardware form to assist in analyzing data. Direct input of data from the measuring point to the computer is often desirable. Other options allow acquisition of data on digital or magnetic tape for varying the method of analysis at a later time. Digital tape printers may be used to acquire data at speeds adequate for the study. Computer analysis normally requires a data conversion from the printed tape to punched cards or magnetic tape.

Each investigation and the types of measurements to be made should be considered in terms of the equipment needed for analysis of the data. Long-term studies may require equipment to scale time to an analyzer operating at electronic speed. Short-term studies may allow connection from terminals to a central computer or directly to a local analyzer programed for the desired information.

2-14. Error Analysis.—Sources of errors in measurements taken during model studies are numerous. Predicting or calculating these errors is most important in establishing the reliability of the studies. Measurements repeated under supposedly identical conditions can yield a variety of answers. The reasons for the variations are many, including alteration of the supposedly identical conditions caused by instrument instability, limitations in investigator judgment, and fluctuations inherent in the random nature of fluid mechanics. Many of these variations may be termed "statistical" and are not under the control of the investigator. Statistical analysis of the variations leads to conclusions on the reproducibility of the measurements. A systematic error, under the control of the investigator, will not be revealed by statistical analysis but must be eliminated through careful study of equipment and methods.

A first source of possible error is in the accuracy of model construction. For example, an improperly built spillway crest causes a systematic error in the depths measured for discharges over the crest. An error of plus or minus 1 millimeter in the model construction causes an error in the depth measurement that may be 50 to 100 times larger when converted to a prototype value. Model inaccuracies also affect pressures measured on a spillway or valve surface. An accurately constructed profile will indicate and control the correct pressures and discharges, and statistical interpretations of the results can be made with confidence. Conversely, inaccurate profiles may indicate satisfactory pressures which are in reality incorrect for the design.

A second source of error comes from the instrumentation. Each measurement system–discharge, depth, pressure, vibration, etc.–should be evaluated for its probable range of error. Primary and secondary standards of length, flow, pressure, etc., should be available for continual checking of measuring instruments. When possible, repetitive measurements should be made to give a sufficient number of samples for statistically determining the probable error.

Statistical methods of analysis are available for analyzing the error of a series of measurements from a model. These methods may be applied to the basic equation governing the related quantities. In many cases, the principal error is the inherent statistical error caused by the natural fluctuations in the rate of occurrence.

CHAPTER 3

Model Design

3-1. **Methods of Analysis.**–Mathematical analyses and experience do not necessarily provide sufficient information to assure satisfactory performance and safety of a costly structure; therefore, model studies are often made to obtain additional knowledge of what to expect when a field structure is put into service.

Dimensionless parameters may be used to classify and generalize experimental results from the models. Experimental data obtained from a flow through a structure at one scale are, under proper conditions, applicable to a different size structure confined by geometrically similar boundaries. Several methods have been developed for obtaining the dimensionless similarity parameters. These methods include: (1) the π theorem, (2) the method of similitude, and (3) the systematic use of differential equations.

The three methods are described in order of increasing utility for problem solving. The π theorem is the least powerful and is normally intermediate in ease of application. Systematic use of differential equations provides the greatest power for finding the solution but is the most complicated to apply. If differential equations are available in reasonably detailed form, the systematic use of these equations should provide a solution superior to the π theorem and similitude methods. Mathematical models having relatively complete differential equations tend to be important as primary means for problem solution.

3-2. **Dimensional Analysis and the π Theorem.**–Buckingham's π theorem [10] is a formal means of arranging dimensional parameters of a physical equation in nondimensional groups or combinations. When properly chosen, dimensionless parameters often will correlate and generalize the results of an investigation. The use of dimensionless parameters reduces the number of required independent variables and the time involved in computing or measuring coordinates.

Variables can be organized into the smallest number of dimensionless groups by the π theorem. A physical equation involves n variables of the form

$$f(a_1, a_2, a_3, ..., a_n) = 0 \qquad (1)$$

The π theorem states that this relationship can be written in terms of m nondimensional parameters

$$f(\pi_1, \pi_2, \pi_3, ..., \pi_m) = 0 \qquad (2)$$

where each π is an independent dimensionless relationship of some of the a variables and m is given by the equation $m = n - k$, in which k is the largest number of variables in the a_1, a_2, ... list that will not combine in dimensionless form.

Conditions which should be fulfilled in using the π theorem are:

(1) The a group must include all the variables of physical significance,
(2) Dimensionless parameters, π's, must contain each of the variables in the a group at least once, and
(3) Dimensions of physical variables must be independent of each other.

Illustrations for applying dimensional analysis are numerous in fluid mechanics textbooks and engineering handbooks. The method offers a means for better understanding the problem, but has limitations. Physical quantities not included in the a listing will not be present in the π parameters and the method does not provide a direct means for finding all the needed variables.

Exceptions can be found in applying the π theorem, and the theorem does not disclose conditions for neglecting π terms. No means are available in the theorem for determining the importance or the informative nature of sets of the dimensionless parameters.

In simple problems, the theorem gives remarkably complete and accurate answers; in complicated problems, it shows the utility of dimensionless groups. The method provides a base for the analysis of units, dimensions, and relationships in a variety of engineering problems [11].

3-3. Method of Similitude.–Geometric similarity requires that all length ratios between model and prototype be the same. Kinematic similarity is a correspondence of motion; in two kinematically similar systems, particle motion will be similar. Dynamic similarity occurs when the ratios of forces are the same in the two systems.

In applying the method of similitude to the design of models for fluid mechanics studies, forces of importance, including the dependent and independent forces, are enumerated by the investigator. These forces are expressed in terms of the problem parameters and variables.

Fluid mechanics commonly involves six forces: inertia, pressure, gravity, viscous shear, surface tension, and elastic compression. Correlating groups of these forces expresses the required similarity of two systems. Two systems having geometric, kinematic, and dynamic similarity have complete similitude.

The correspondence between a hydraulic model and its prototype is limited because the similitude for one or more forces is usually incomplete. Geometric similarity is independent of motion, but force similarity depends on selection of geometric scale. Scale selection should be made to satisfy the predominant force and to make negligible the remaining forces.

The similitude relationship of the six forces normally encountered in fluid mechanics studies is derived from Newton's second law of motion:

$$F_i = Ma = \text{vector sum} \quad F_p + F_g + F_v + F_t + F_e \qquad (3)$$

where

Symbol		Force	Dimension
F_i	=	Inertia	$\rho V^2 L^2$
F_p	=	Pressure	$\Delta p L^2$
F_g	=	Gravity	$\rho g L^3$
F_v	=	Viscous shear	$\mu V L$
F_t	=	Surface tension	$S L$
F_e	=	Elastic compression	$E L^2$

ρ is density, L is length, V is velocity, Δp is pressure drop, μ is dynamic viscosity, E is fluid bulk modulus, S is coefficient of surface tension, and g is local acceleration due to gravity.

For dynamic similitude, the ratio of the inertia forces must equal the ratio of the vector sum of the active forces:

$$\frac{(F_i)_m}{(F_i)_p} = \frac{(F_p + F_g + F_v + F_t + F_e)_m}{(F_p + F_g + F_v + F_t + F_e)_p} \tag{4}$$

Complete similitude occurs if

$$\frac{(F_i)_m}{(F_i)_p} = \frac{(F_p)_m}{(F_p)_p} = \frac{(F_g)_m}{(F_g)_p} = \frac{(F_v)_m}{(F_v)_p} = \frac{(F_t)_m}{(F_t)_p} = \frac{(F_e)_m}{(F_e)_p} \tag{5}$$

No model fluid has the viscosity, surface tension, and elastic characteristics to satisfy the conditions of equation (5). Nevertheless, application of equation (5) is not difficult because, in the majority of hydraulic model studies, neglecting the effects of surface tension and elastic forces produces only minor errors. Models can closely simulate fluid motion if either gravity or viscous forces predominate; correction for the effects of other forces is required occasionally.

Dimensionless parameters for each of the five forces related to the inertia force can be prepared for use in fluid mechanics studies. A practical basis for similitude in models where gravitational forces predominate is the equating of inertia/gravity force ratios of model and prototype. In dimensional terms,

$$\left(\frac{F_i}{F_g}\right)_m = \left(\frac{F_i}{F_g}\right)_p$$

$$\left(\frac{\rho V^2 L^2}{\rho g L^3}\right)_m = \left(\frac{\rho V^2 L^2}{\rho g L^3}\right)_p$$

$$\left(\frac{V^2}{gL}\right)_m = \left(\frac{V^2}{gL}\right)_p$$

$$\frac{(V_m/V_p)^2}{(g_m/g_p)(L_m/L_p)} = 1$$

$$\frac{V_r^{\,2}}{g_r L_r} = 1 \quad \text{or} \quad \frac{V_r}{\sqrt{g_r L_r}} = 1 \tag{6}$$

$$\frac{V}{\sqrt{gL}} = \text{Froude number} \tag{7}$$

Equating the inertia/viscous force ratios of model and prototype forms a basis for similitude in studies having a predominant viscous force. In dimensional form,

$$\left(\frac{F_i}{F_v}\right)_m = \left(\frac{F_i}{F_v}\right)_p$$

$$\left(\frac{\rho V^2 L^2}{\mu VL}\right)_m = \left(\frac{\rho V^2 L^2}{\mu VL}\right)_p$$

$$\left(\frac{\rho VL}{\mu}\right)_m = \left(\frac{\rho VL}{\mu}\right)_p$$

$$\frac{\rho_r V_r L_r}{\mu_r} = 1 \tag{8}$$

$$\frac{\rho VL}{\mu} = \frac{VL}{\nu} = \text{Reynolds number} \tag{9}$$

where

μ = dynamic viscosity
ν = kinematic viscosity

Similarly, other nondimensional parameters useful in fluid mechanics studies can be derived:

When surface tension predominates,

$$\frac{F_i}{F_t} = \frac{\rho V^2 L}{S} = \text{Weber number} \tag{10}$$

When elastic compression predominates,

$$\frac{F_i}{F_e} = \frac{\rho V^2}{E} = \frac{V^2}{C^2} = \text{Cauchy number} \tag{11}$$

(where C = sonic velocity)

or

$$\frac{F_i}{F_e} = \sqrt{\frac{V^2}{C^2}} = \frac{V}{C} = \text{Mach number} \tag{12}$$

The Mach number is used in studies that use air as the flowing fluid.

For complete similarity of tension or elastic forces, the Weber or Cauchy numbers, respectively, are equal in the model and the prototype [12].

3–4. **Differential Equations.**–Differential equations written for an elemental volume of fluid are the most complete and detailed equations governing model design. These equations must describe the physical and mathematical conditions of the problem. This description becomes a mathematical model limited to what are considered to be the important physical aspects. Those aspects excluded determine the degree of uncertainty of the model and the inaccuracies acceptable for given circumstances.

In applying this method, the investigator must clearly understand the physical aspects and the limitations of mathematical models. The equations of the model are carefully defined in form and magnitude by the dependent nondimensional variables. Boundary conditions are established to be approximately equal to, but not greater than, 1 for the nondimensional variables in the same domain. The independent nondimensional parameters are defined in the domain of the dependent variables and for approximately the same boundary conditions.

The differential equation containing dependent and independent nondimensional parameters is made dimensionless. A coefficient from one term of the equation is used to divide through the equation, term by term, in a process called normalization. The usefulness of the remaining dimensionless π groups depends on the care and insight in choosing the form of the parameters.

Consider steady flow in a pipe (fig. 3-1). Appropriate relationships between model and prototype frictional effects are desired. The first step is to obtain the differential equation describing the motion. For instance, the forces on an incremental element of the flow could be equated (fig. 3-2).

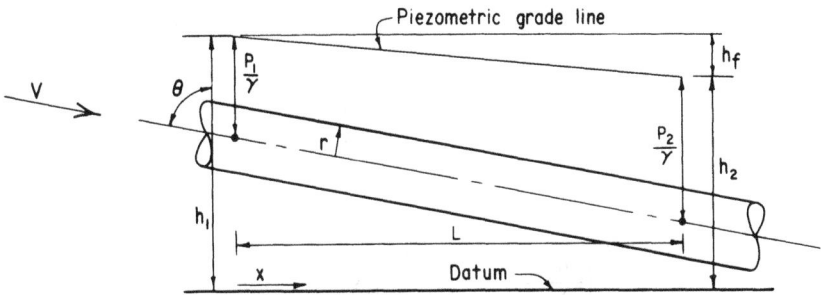

Figure 3-1.–Pressure variation in a pipe. 103-D-1709.

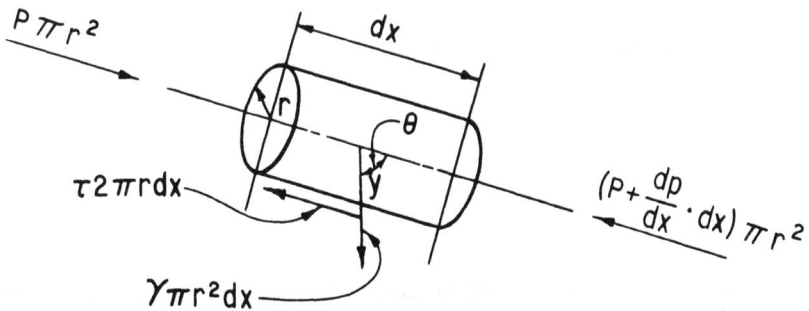

Figure 3-2.–Pressure and shear on an incremental element of flow. 103-D-1710.

Setting the sum of the forces in the longitudinal direction equal to zero gives:

$$\underbrace{P\pi r^2 - (P + \frac{dp}{dx} \cdot dx)\pi r^2}_{\text{pressure forces}} \underbrace{- \tau 2\pi r dx}_{\substack{\text{shear} \\ \text{force}}} \underbrace{+ \gamma \pi r^2 dx \cdot \cos\theta}_{\text{body force}} = 0 \quad (13)$$

where

$$P = \text{pressure}$$
$$\tau = \text{shear stress}$$
$$\gamma = \text{specific force}$$

Dividing by πr^2 and simplifying gives

$$-\frac{dp}{dx} \cdot dx - \frac{2\tau dx}{r} + \gamma dx \cdot \cos\theta = 0 \quad (14)$$

or

$$\left(\frac{dp}{dx} - \gamma \cdot \cos\theta\right) dx = -\frac{2\tau}{r} dx$$

But the piezometric grade line is defined by

$$\frac{1}{\gamma} \frac{dp}{dx} \cdot dx - \cos\theta \cdot dx = dh$$

By combining these two equations, the slope of the piezometric grade line is determined to be:

$$\frac{dh}{dx} = -\frac{2\tau}{\gamma r} \quad (15)$$

Assuming the fluid is Newtonian[1] and no secondary flows exist,

$$\tau = \mu \frac{du}{dr}$$

Thus,
$$\frac{dh}{dx} = -\frac{2\mu}{\gamma r}\frac{du}{dr}$$

Since
$$\mu = \rho\nu \text{ and } \gamma = \rho g,$$

then
$$\frac{dh}{dx} = -\frac{2\nu}{gr}\frac{du}{dr} \tag{16}$$

The boundary conditions are such that

x varies between 0 and L
r varies between 0 and R
u varies between 0 and $\overline{U}$
$\overline{U} =$ maximum velocity
h varies between h_1 and h_2

The second step is to form dimensionless ratios varying between 0 and 1. This can be done by defining

$\overline{x} = x/L$
$\overline{r} = r/R$
$\overline{u} = u/\overline{U}$
$\overline{h} = (h/h_1 - h_2) = h/h_f$
(h_f is loss in piezometric head caused by friction)

---

[1] A Newtonian fluid is defined as a fluid that has a linear stress versus rate-of-strain relationship, with zero rate-of-strain at zero shear stress.

Substituting these values into the differential equation gives:

$$\frac{h_f\,d\bar{h}}{L\,d\bar{x}} = \frac{-2\nu\,\bar{U}\,d\bar{u}}{gR^2\,\bar{r}\,d\bar{r}}$$

Substituting $D/2$ for R,

$$\frac{h_f\,d\bar{h}}{L\,d\bar{x}} = \left(\frac{-2\nu}{g(D/2)}\right)\left(\frac{\bar{U}}{(D/2)}\right)\left(\frac{1}{\bar{r}}\frac{d\bar{u}}{d\bar{r}}\right) \tag{17}$$

The third step is to divide each term of the equation by the coefficients of one term. Thus, dividing by h_f/L and grouping gives:

$$\frac{d\bar{h}}{d\bar{x}} = -16\left(\frac{\bar{U}^2/2gD}{h_f/L}\right)\left(\frac{\nu}{\bar{U}D}\right)\frac{1}{\bar{r}}\frac{d\bar{u}}{d\bar{r}} \tag{18}$$

The term $\dfrac{n_f/L}{\bar{U}^2/2gD}$ is the same as f (friction coefficient) in the

Darcy-Weisbach formula and $\bar{U}D/\nu$ is a Reynolds number.

Thus, a functional relationship between r and the Reynolds number should be expected for a given pipe geometry, that is, for the same wall surface. If the wall surface is varied in roughness, the effect can be accounted for by the addition of another geometric term, ϵ/D.

A complete and detailed differential equation having proper boundary conditions will contain a sufficient set of dimensionless parameters for modeling procedures. Choices of parameters provide results of varying utility. Unfortunately, the most useful are not known in advance of formulating the parameters.

As evidenced in the dimensional analysis and similitude methods, decreasing the number of parameters and conditions in the governing equations may increase the ease of establishing similarity. Thus, critical study of the equations and parameters should be made to provide a minimum of conditions to be placed on the similitude [13].

46

An investigator must recognize the complexities of a problem and select a method appropriate to the solution. Dimensional analysis, similitude, or differential equation methods may be used in the design of a model. The use of governing equations is a powerful method of establishing similarity, but is also the most complicated to apply. Differential equations are available in fairly detailed form for many conditions encountered in the study of fluid mechanics, and this method produces the best opportunity for understanding the problem and gaining the maximum of information in the solution.

3-5. **Application to Physical Models.**–Satisfying more than one parameter in a model investigation requires that the physical properties of the testing fluid be variable over rather broad limits. For example, to satisfy the Froude and Reynolds laws simultaneously, it would be necessary that:

$$\frac{V_r}{\sqrt{g_r L_r}} = \frac{L_r V_r}{\nu_r}$$

Because the acceleration due to gravity is essentially the same in model and prototype, that is, $g_r = 1$, the kinematic viscosity ratio would have to be related to the length scale as follows:

$$\nu_r = L_r^{3/2}$$

Satisfying this criterion for a hydraulic structure model with a length scale of 1 to 25 would require a test fluid having a kinematic viscosity 1/125 that of water. Such a fluid is not available. The physical properties of a practical testing fluid are such that only the dominant parameter can be modeled in a given study.

The Froude parameter, equation (7), is used most frequently in hydraulic problems involving turbulent free-surface flow. Conditions assumed in formulating the parameter are essentially realized in the case of turbulent flow with a free water surface because the effects of gravity outweigh those of viscosity and surface tension. However, when the viscous and surface tension forces are neglected, every effort should be made to minimize them by using large models and smooth boundaries. Normally, when the Reynolds number of the model is greater than 1×10^4, and depth of flow is substituted for L in the Reynolds number, the viscous forces are relatively unimportant. Compensating adjustments to slope or boundary roughness can be made if viscous effects cannot be ignored. These adjustments will be discussed later in connection with the type of model to which each is applicable.

Steady flow in a pressure conduit, or flow around a deeply submerged body, approximates the conditions assumed in formulating the Reynolds parameter, equation (9). With no free surface directly involved in the flow pattern, and with steady flow, the forces of surface tension and elasticity are eliminated. Further, the gravity forces are balanced and therefore do not affect the flow pattern. The Reynolds number in the model can rarely equal that of the prototype, but fortunately this is not necessary. If the Reynolds number for a model with smooth boundaries exceeds 1×10^6 for all pertinent flows, test data will be satisfactory even if the model Reynolds number does not equal that of the prototype, because boundary resistance coefficients do not change appreciably for Reynolds numbers above this value. This can be shown when the dimensionless friction factor, f, in the Darcy-Weisbach formula, $h_f = f(L/D)/(V^2/2 \, g)$, is plotted with respect to the Reynolds number for flow in pipes with various relative roughness coefficients, ϵ/D. For values of ϵ/D less than 0.001, the friction factor becomes nearly constant when the Reynolds number is greater than 1×10^6 and the ratio of resistance forces, model to prototype, depends only on the ratio of relative roughnesses. To obtain a satisfactorily large Reynolds number in a small model, the velocity is often increased by using larger-than-scaled pressure heads to provide similar boundary resistance forces.

When surface tension is the predominating factor in a study, the other three forces are usually insignificant and the Weber law, equation (10), applies. Surface tension is manifest in capillary waves on a water surface, capillarity in manometer tubes, capillary wetting action in filter beds and earth dams, tendency of a jet from an orifice to assume a circular cross section regardless of shape of orifice, and in flow over weirs at low heads. Surface tension seldom plays an important role in model testing because of the choice of selecting the scale and mode of model operation [14].

Except for cases of unsteady flow, especially waterhammer problems, similitude based on the Mach number, equation (12), has little application in hydraulic model testing. Since most waterhammer problems yield readily to analytical methods, model studies are rarely needed. On the other hand, aerodynamic investigations of compressible flow led to an extensive use of the

Mach law to deal with problems involving the flow of gases at velocities exceeding the speed of sound.

After deciding on the dominant dimensionless parameter, the similitude ratios can be developed for a model study. For open-channel flow where the Froude law applies, the model/prototype relationships are obtained directly from equation (6):

$$\frac{V_r}{\sqrt{g_r L_r}} = 1 \tag{6}$$

or $\quad V_r = \sqrt{g_r L_r}$

Since $g = \gamma/\rho$, the velocity ratio can also be written

$$V_r = \left(\frac{\gamma L}{\rho}\right)_r^{1/2}$$

Other kinematic relationships are as follows:

time ratio, $\quad t_r = \dfrac{L_r}{V_r} = \dfrac{L_r}{\left(\dfrac{\gamma L}{\rho}\right)_r^{1/2}} = \left(L\dfrac{\rho}{\gamma}\right)_r^{1/2}$

acceleration ratio, $\quad a_r = \dfrac{L_r}{t_r^2} = \dfrac{L_r}{\left(\dfrac{L_r}{V_r}\right)^2} = \dfrac{V_r^2}{L_r} = \dfrac{\left[\left(\dfrac{\gamma L}{\rho}\right)_r^{1/2}\right]^2}{L_r} = \left(\dfrac{\gamma}{\rho}\right)_r$

discharge ratio, $\quad Q_r = \dfrac{L_r^3}{t_r} = \dfrac{L_r^3}{\dfrac{L_r}{V_r}} = L_r^2 V_r = L_r^2 \left(\dfrac{\gamma L}{\rho}\right)_r^{1/2} = L_r^{5/2} \left(\dfrac{\gamma}{\rho}\right)_r^{1/2}$

The dynamic relationships are obtained in the same manner: Since the mass ratio $M_r = L_r^3 \rho_r$,

the force ratio, $\quad F_r = M_r a_r = (L^3 \rho)_r \left(\dfrac{\gamma}{\rho}\right)_r = (L^3 \gamma)_r$

When the fluid in the model is the same as that in the prototype, both γ_r and ρ_r are unity.

In closed-conduit problems where the Reynolds law applies, another set of similitude ratios may be developed from equation (8) as follows:

$$\frac{\rho_r V_r L_r}{\mu_r} = 1 \tag{8}$$

velocity ratio, $\qquad V_r = \left(\dfrac{\mu}{L\rho}\right)_r$

time ratio, $\qquad t_r = \dfrac{L_r}{V_r} = \dfrac{L_r}{(\mu/L\rho)_r} = \left(\dfrac{L^2 \rho}{\mu}\right)_r$

acceleration ratio, $\qquad a_r = \dfrac{L_r}{t_r^2} = \left(\dfrac{\mu^2}{L^3 \rho^2}\right)_r$

and discharge ratio, $\qquad Q_r = \dfrac{L_r^3}{t_r} = \left(\dfrac{L\mu}{\rho}\right)_r$

Table 3-1 shows the relationships for flow in which gravity and viscous forces predominate. The viscous force ratios can only be used for problems of lubrication and a very limited number of hydraulic structure and equipment studies in which the prototype Reynolds number is low. For example, in a hydraulic model having a scale ratio of 1:20, a velocity ratio of 20:1 would be required to

Table 3-1.—*Scale ratios for investigations in which gravity and viscous forces predominate*

Characteristic	Dimension	Scale ratios for the laws of	
		Froude	Reynolds
Geometric properties			
Length	L	L_r	L_r
Area	L^2	L_r^2	L_r^2
Volume	L^3	L_r^3	L_r^3
Kinematic properties			
Time	t	$\left[\dfrac{L\rho}{\gamma}\right]_r^{1/2}$	$\left[\dfrac{L^2\rho}{\mu}\right]_r$
Velocity	Lt^{-1}	$\left[\dfrac{L\gamma}{\rho}\right]_r^{1/2}$	$\left[\dfrac{\mu}{L\rho}\right]_r$
Acceleration	Lt^{-2}	$\left[\dfrac{\gamma}{\rho}\right]_r$	$\left[\dfrac{\mu^2}{\rho^2 L^3}\right]_r$
Discharge	$L^3 t^{-1}$	$\left[L^{5/2}\left(\dfrac{\gamma}{\rho}\right)^{1/2}\right]_r$	$\left[\dfrac{L\mu}{\rho}\right]_r$
Dynamic properties			
Mass	M	$(L^3\rho)_r$	$(L^3\rho)_r$
Force	MLt^{-2}	$(L^3\gamma)_r$	$\left[\dfrac{\mu^2}{\rho}\right]_r$
Density	ML^{-3}	ρ_r	ρ_r
Specific weight	$ML^{-2}t^{-2}$	γ_r	$\left[\dfrac{\mu^2}{L^3\rho}\right]_r$
Pressure intensity	$ML^{-1}t^{-2}$	$(L\gamma)_r$	$\left[\dfrac{\mu^2}{L^2\rho}\right]_r$
Impulse and momentum	MLt^{-1}	$[L^{7/2}(\rho\gamma)^{1/2}]_r$	$(L^2\mu)_r$
Energy and work	ML^2t^{-2}	$(L^4\gamma)_r$	$\left[\dfrac{L\mu^2}{\rho}\right]_r$
Power	ML^2t^{-3}	$\left[\dfrac{L^{7/2}\gamma^{3/2}}{\rho^{1/2}}\right]_r$	$\left[\dfrac{\mu^3}{L\rho^2}\right]_r$

satisfy the Reynolds law. The prototype velocity would normally exceed 1.5 meters per second and for similitude would require a model velocity of over 30 meters per second. Such velocities are beyond the capability of most hydraulic laboratories, but an approximate similitude can be obtained by operating the model at Reynolds numbers in the order of 1×10^6.

The similitude relationships governed by the Weber and Cauchy parameters can be developed from equations (10) and (11), respectively.

3–6. **Distorted Models.**–Models of river channels, estuaries, floodways, and canal structures are often distorted geometrically when surface tension or tractive forces are not properly scaled in the model. The undistorted channels would have insufficient cross section to obtain representative flow conditions unless the model were extremely large. In a distorted model, the scale ratios may be different for length, width, and depth. Conditions for distortion vary greatly from one model to another. When a distorted model is used instead of a true model, the distortion is planned to accomplish a definite objective, e.g., scaled depth, scaled velocity, or surface roughness. The results obtained from such a model are limited to this objective.

Advantages of distorted models are:

- Sufficient tractive force can be developed to produce bedload movement with a reasonably small model and available model sediment.
- Water surface slopes are exaggerated and therefore easier to determine.
- The width and length of the model can be held within economical limits for the required depth.
- Operation is simplified by use of a smaller model.
- Turbulent prototype flow can be modeled.

Disadvantages are:

- Velocities may not be correctly reproduced in magnitude and direction.
- Some of the flow details are not correctly reproduced.
- Slopes of cuts and fills are often too steep to be molded in sand or erodible material.
- There is an unfavorable psychological effect on the observer who views distorted models.
- Boundary roughness may need to be distorted to produce similarity.

The similitude ratios for distorted models based on the Froude law are:

$$\text{Horizontal length ratio} = L_r$$
$$\text{Depth ratio} = D_r$$
$$\text{Horizontal area ratio} = L_r^2$$
$$\text{Vertical area ratio} = L_r D_r$$
$$\text{Slope ratio} = D_r / L_r$$
$$\text{Velocity ratio} = D_r^{1/2}$$

In problems requiring the use of distorted models, resistance often plays a significant part. Then the ratio of model-to-prototype resistance must be made the same as the ratio of gravity forces. The resistance may be related to the velocity by means of the Manning equation:

$$V = \frac{R^{2/3} S^{1/2}}{n}$$

where

$$S = \text{the resistance slope}$$
$$R = \text{hydraulic radius}$$
$$n = \text{roughness coefficient}$$

To make the resistance slope equal to the channel slope, the following relationship must be satisfied:

$$S_r = \frac{D_r}{L_r} = \frac{n_r^2 \, V_r^2}{R_r^{4/3}} = \frac{n_r^2 \, D_r}{R_r^{4/3}}$$

Then

$$n_r = \frac{R_r^{2/3}}{L_r^{1/2}}$$

or

$$n_m = n_p \left(\frac{R_r^{2/3}}{L_r^{1/2}} \right)$$

The value of the hydraulic radius ratio, R_r, does not bear fixed relations to the length and depth scales, but varies with the shape of the cross section. Then, for any given reach, the R_r and the L_r are known, and n_r can be computed and applied to the prototype n to obtain the required value of n for the model. Usually the required value of n in the model (distorted or undistorted) must be obtained by artificially roughening the model channels or by adjusting the slope for a given roughness. Unless basic data on the values of n for different types of roughness are available, the appropriate roughness must be obtained by experiment. The model with roughness adjusted for a particular depth will yield dependable results for flow at or near that depth. If studies include several depths, model roughness should be adjusted to give an average friction that is approximately right for each depth, or the roughness may be varied with depth for a closer approximation at all depths. The fact should not be overlooked, however, that the Manning n is strictly a roughness characteristic and that the Manning equation applies only at sufficiently high values of Reynolds number where resistance forces are proportional to the square of the velocity.

Generally, distorted models are not adapted to situations where curvatures in the water surface are involved. For example, the flow through spillways or between bridge piers will not be correctly represented in a distorted river model.

An important precaution in testing distorted river models is to be sure that flow in the model is of the same type as in the prototype. Flow is turbulent in rivers and, as viscous forces predominate, it appears that a criterion for model design would be the Reynolds parameter or some variation of the factors which constitute this number. A practice in some American laboratories has been to select the scales in the model to provide a Reynolds number of not less than 2500, using the hydraulic radius for length. The Waterways Experiment Station of the Corps of Engineers has used the criterion $VR \geq 0.02$ to ensure turbulent flow, where V is average velocity in the channel and R is hydraulic radius. This criterion corresponds to a limiting Reynolds number of 1800 at a temperature of 19 °C.

The SOGREAH[1] laboratory at Grenoble, France, tries to ensure a turbulent flow that may be described as the "rough-wall type" or "rough-turbulent type." This type of flow is required in models for all cases except for very sluggish rivers. In the usual application, the Reynolds number is not a completely satisfactory parameter for open-channel flow, nor an adequate criterion for river model turbulence. In place of the Reynolds number, SOGREAH proposed

---

[1] Societe Grenobloise, d'Etudes et d' Applications Hydrauliques.

a "roughness Reynolds number" or a "Karman number," in honor of Professor Theodore von Karman of the California Institute of Technology [15]:

Karman number, $$\overline{K} = \frac{kV_*}{\nu} \tag{19}$$

where k is a roughness parameter that, for the usual range of slopes, is proportional to the sand grain size, and

$$V_* = \text{shear velocity} = \sqrt{RSg}$$

In proportioning models to obtain the proper degree of turbulence, a Karman number not less than 100 is recommended.

CHAPTER 4

Laboratory Studies

4-1. **General.** –If the application of established design procedures and available information fails to provide a solution to a hydraulic problem, then a laboratory study should be made. In this book, a laboratory study is defined as any investigation of a hydraulic problem carried out by the laboratory staff. An investigation may consist of analytical studies, laboratory experimentation, field testing, or a combination of these.

The hydraulic problems should be thoroughly examined and discussed by design and laboratory engineers. The laboratory engineer thus acquires a familiarity with design considerations and construction limitations, while the designer learns the potential as well as the limits of a laboratory study of the problem. A mutual selection can then be made of the appropriate study technique that will assure development of a practical solution to the problem.

Defining the problems involved in a study is not always possible before construction of a model. If this is the case, selection of a model scale and construction of the model must then proceed on the basis of experience with similar studies. Discussion of the study may indicate uncertainty with respect to general arrangement of the components of the structure flow passages rather than specific sources of trouble. An experienced hydraulic engineer can generally detect the need for design changes by observing the model in operation. Various types of hydraulic structures and machinery are discussed in this book to provide performance criteria that will be helpful in appraising the hydraulic behavior of a model or prototype.

Before constructing a model, account must be taken of limitations imposed by funds, time, and availability of personnel in relation to the size or type of model. The desired accuracy of the final results will also materially affect the type and extent of the investigation.

For example, a small spillway model may be adequate to indicate upstream or downstream riverflow conditions or erosion tendencies, while a study of discharge coefficients for gates or an overflow crest may require a larger and more detailed model. Limitations of space or pumping facilities are additional factors controlling model size. For example, a large model of a river system can be tested satisfactorily only where large floor or outdoor areas are available. However, constructing and testing the model in sections may be satisfactory, provided a method can be devised for correlating the results from each section.

Although a model may be carefully designed and constructed, it will not necessarily provide a direct solution to the problem. Information obtained must be interpreted using not only a knowledge of basic mechanics and hydraulics, but experience as well. Although a model serves to demonstrate fundamental principles that apply equally to the prototype, complete similitude is seldom satisfied. Thus, any direct translation of results must be applied with understanding and restraint.

The laboratory engineer should be innovative in approaching the design and operation of a model. Time taken in the design to simplify construction for possible later alterations reduces the model costs. In designing the model, careful consideration of the type of data and method of analysis eases the interpretation of results as the investigation progresses.

4-2. **Prototype Information Required.**–Information for use in planning a model should include drawings depicting an overall plan and cross sections of the proposed or existing prototype structure. Sufficient detail should be given to determine the shapes and characteristics of all surfaces over which the flow will pass. Topography of the site and surrounding area, results of foundation test boring, and details of other related structures (particularly the hydraulic features) may be necessary for successful model plans. Knowledge of the types and condition of materials composing the riverbed and canyon walls may be useful in establishing the rugosity of the model channel. A reliable curve showing water stage and discharge for the river downstream from the structure is important. In analyzing river or estuary problems, historical data, such as discharge, water stage, tide salinity, sediment-carrying capacity, cross sections of channels, natural hydraulic gradient controls, and depths, are essential. For spillways and outlet works, a complete description of proposed operating conditions and the range of discharges to be expected should be available.

Problems relating to a model study of the structure should be discussed with the designers of the structure or other responsible persons familiar with the project. Full knowledge of local and general design conditions affecting hydraulic performance will often save considerable work and time. The importance of becoming thoroughly familiar with design requirements early in the study cannot be overemphasized.

4-3. Design and Construction of Model.–(a) *Scale.*–Success in achieving the desired results from a laboratory study in the least time and with the least expense depends largely on the design of the model. The first and most important step in the design is the careful selection of a model scale. In general, large rather than small models should be built, as permitted by available space and water supply. Small models (for example, scale ratio $L_r = 1{:}100$) may be used advantageously in preliminary studies to give direction to the primary investigation. Although measurement accuracy may be limited, information can be obtained from a small model that is applicable to a larger model. For maximum similarity between the model and prototype, the model for the primary study should be made as large as possible, taking into consideration its cost and the derived benefits. A large model ($L_r = 1{:}5$) is more useful than a small one, and improves the accuracy of measurements, but at some point the cost and difficulty of operation will offset the advantage of large size.

The following scale ratios have been used successfully in Bureau of Reclamation model studies and may be useful as a guide. Spillways for large dams have been constructed on scale ratios ranging from 1:30 to 1:100. The model for a medium-size spillway should not be smaller than that determined by a scale ratio of 1:60. Outlet works having gates and valves are constructed to scale ratios ranging from 1:5 to 1:30. Canal structures, such as chutes and drops, are usually constructed on scale ratios ranging from 1:3 to 1:20. The range of horizontal scale ratios for river models is usually between 1:100 and 1:1000, and the vertical scale ratio for distorted river models, between 1:20 and 1:100.

Minimum model sizes have been established for certain types of studies. Models of valves, gates, and conduits should have flow passages at least 100 millimeters across so that, for the heads and discharges normally available in the laboratory, turbulent flow is produced in the pipeline and valve. The scale of an outlet model is often determined from this minimum diameter or width. In models of canal structures, the bottom width of channels should exceed 100 millimeters so as to provide turbulent flow. To reduce the effects of viscosity and surface tension, spillway models should be scaled to provide flow depths over the crest of at least 75 millimeters for the design normal operating range. Models designed to dimensions smaller than these are difficult to build to the required accuracy. Model valves and gates for use in development work or for rating of prototype control devices should be at least 150 millimeters in diameter or width. Time and expense may be saved by choosing the scale for tunnel or conduit models to accommodate a standard pipe size or sections of pipe already on hand. Also, use can be made of parts of previously tested models.

(b) *Materials.*–A model need not be made of the same materials as the prototype. If surfaces over which the water flows are reproduced in shape, and the roughness of the surfaces is approximately to scale (normally should be smoother in the model than in the prototype), the model will usually be satisfactory. Figures 4-1 and 4-2 show typical model construction. Materials for models are selected according to the availability, cost, and precision of construction necessary in the particular part of the structure.

Figure 4-1.–Model of morning glory spillway under construction. P801-D-79229

60

Contours and supports in reservoir section of model. P801-D-74588

Wire lath and cement-sand mortar overlay adjacent to dam. P801-D-79228

Downstream channel before cement-sand mortar and placing of movable-bed material. P801-D-74587

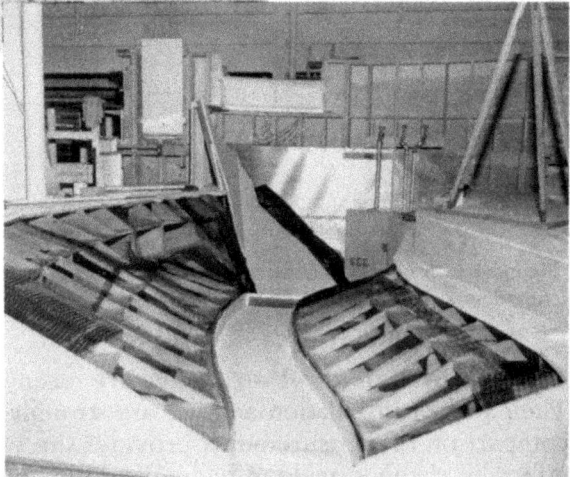

Figure 4-2.–Stages in constructing an arch dam spillway model.

Boxes enclosing the models may be made of metal, wood with sheet metal lining, or waterproof plyboard with joints sealed by plastic tape and mastic (fig. 4-3). Irregular flow surfaces of models are often formed of cement-sand mortar placed over wire lath. Models of tunnels or closed conduits are constructed of sheet metal or transparent sheet plastic. Sheet metal is less expensive and can be used where observation of the flow is not necessary. Valves and gates of the smaller sizes may be fabricated of plastic; however, for development work where high heads are involved, they are constructed of brass. Models of canal structures are constructed of painted wood, wood covered with sheet metal, or plyboard impregnated with a waterproof coating.

Figure 4-3.–Model box construction of waterproof plyboard, plastic tape, and mastic-sealed joints. P801-D-79227

(c) *Instruments.*–Instruments are essential in model testing. Their proper installation and use are strongly emphasized because comparison of measurements provides the basis for studying the effects of changes made in hydraulic designs. Provisions should be made for instrumentation while the model is still in the design stage since instrument connections and piezometers are relatively easy to install early in the construction, but very difficult after the model is completed. Piezometers are relatively inexpensive and should be provided in generous numbers to adequately define the critical pressure areas.

(d) *Model Drawings.*–Accurate and easily understood model drawings prevent time-consuming errors in construction. Drawings should contain sufficient information to allow the building of a model conforming to the design specifications for the structure. Figure 4-4 shows an overall drawing of an earth dam spillway model. Dimensions could be added giving the size of the boxes and location of various parts of the structure. Details are usually shown on other drawings. Important parts which affect the performance of the structure, or special features that are new in concept, should be detailed completely. Having modelmakers who are experienced in hydraulic model construction reduces the need for detailed drawings, and many of the construction methods may be left to the shop manager.

Working drawings that include all information necessary for installation of instruments, such as piezometers, pressure transducers, velocity measuring devices, etc., make extensive changes unnecessary after model completion.

(e) *Construction.*–Because models must be constructed with a high degree of accuracy, craftsmen should be carefully trained to perform their work to close tolerances. Tolerances are particularly close in critical areas such as spillway crests and model valves. Greatest accuracy should be maintained where there will be rapid changes in direction of flow and where high velocities will prevail. The stage of the testing program will affect the type and accuracy of construction. In early tests, when many schemes are tried for feasibility, the construction need not be as carefully performed as later when the data acquired are to be used for constructing and operating the prototype structure.

Many work-reducing methods in constructing and testing models are learned from experience. One such method is that initial construction of the model should allow for considerable modification with a minimum of rebuilding. For instance, a model with a spillway stilling basin set on or close to the laboratory floor will require extensive rebuilding if the basin floor has to be lowered; thus, the original plan should set the basin high enough to eliminate the need for major rebuilding if modifications become necessary.

Viewing the water as it flows through various parts of a model may be very important. For example, flow may tend to form waves

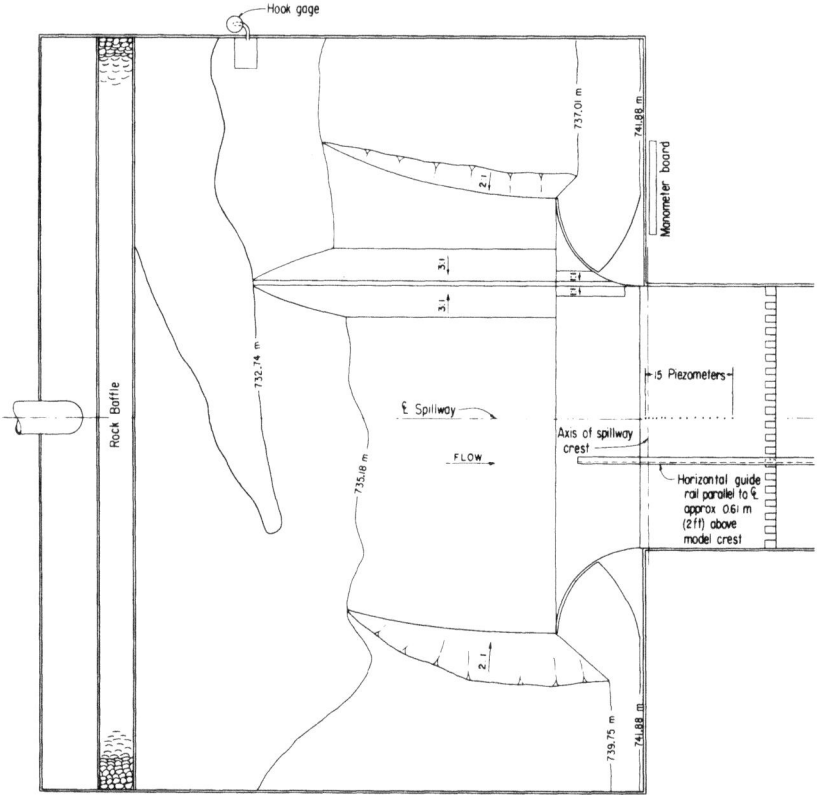

PLAN

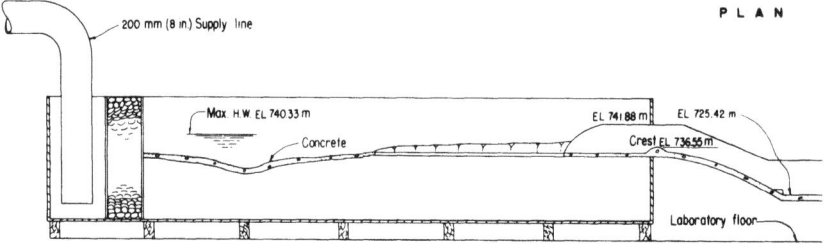

SECTION ON ₵

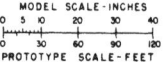

DICKINSON DAM SPILLWAY
MODEL LAYOUT
1:36 MODEL

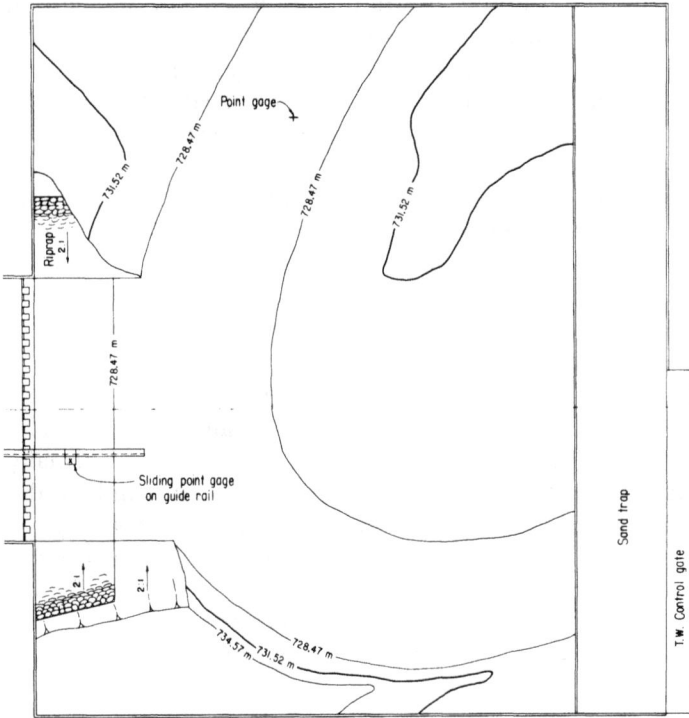

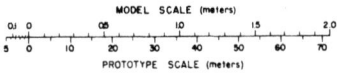

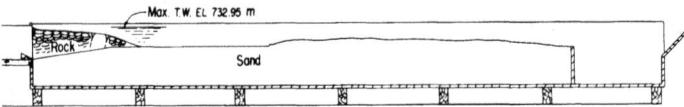

Figure 4-4.–Spillway model drawing. 103-D-1711.

and close a spillway tunnel, thus producing undesirable pressure instead of free-surface flow. Observation of the size and location of the waves is often desirable in deciding upon model changes. Visibility for photographic recording may require clear plastic material for windows in studies of stratified flow, energy dissipation, outlet works, cavitation, etc. A part of the carpentry support shop should be made responsible for understanding and using plastic materials for model construction. The engineer, in cooperation with craftsmen, can design and use clear and opaque plastic materials for models.

A head box and a tail box are usually built for each model. The head box serves as the reservoir upstream from the structure and ensures calm approach conditions. The box must be watertight for accurate discharge measurements through the model, and it usually contains a head gage for measuring the elevation of the water surface in the reservoir.

The tail box at the downstream end of the model (fig. 4-4) contains a gate or other device used to vary the tailwater level. The gate may automatically adjust the tailwater depth but is normally adjusted manually. Sand or gravel placed in the tail box aids in the study of energy dissipation by forming scour patterns. A staff gage or pressure tap in the box indicates the water stage in the river downstream from the structure.

The size of the model box is designed to contain the important flow features of the structure. Parts of the abutment adjacent to an earth dam and spillway and the slope of the dam extending 100 millimeters or so above maximum water surface will normally be adequate to ensure a satisfactory flow approaching a control (fig. 4-1). A complete model may be necessary for studies in which the major portion of the dam controls the flow paths to outlets from the reservoir.

In building a flow channel, contours are scaled from area maps and drawn on wrapping paper to model size. The scaled contours are then transferred to thin lumber and the lumber is cut into contour shapes. With the wrapping paper pattern placed on the model floor in the proper position relative to the dam or control structure, the wooden contours are positioned above the appropriate contours on the pattern, making an allowance for an overlay of wire lath screen and cement-sand mortar, and nailed in place. The allowance for the overlay is a nominal 20 millimeters, either below the elevation or behind the horizontal position. Unless the contour lies in a critical part of a flow path, extreme accuracy of placement is not necessary. Contour construction must be accurate adjacent to spillway entrances, energy dissipators, and other parts of the model where flow distribution and velocity affect the discharge capacity, pressures, and erosion of a channel. Elevations in these critical areas may be established by using an engineers' level and placing nails or pegs on the contours to the finished height of the cement-sand mortar.

Contours may also be formed using sand or small gravel. Profiles are shaped with scaled cross-sectional templates, and a coating of cement-sand mortar is used to finish the flow surface of the channel. This type of contouring is adaptable to shallow, wide channels where slopes are less than the angle of repose of the material.

Sand or gravel is placed in the mortar-finished channels to represent loose or movable river material. Models of river systems may be constructed of cement-sand mortar when a fixed bed is desired, or entirely of sand or granulated plastic materials when sediment movement is part of the study.

For calming the flow of water as it enters the model reservoir, rock baffles upstream from the control structure are provided. They are usually constructed of wire screen (13-mm mesh) over a 2- by 4-inch (50- by 100-mm, nominal) lumber frame and filled with gravel ranging from 19 to 38 millimeters in size, as shown on the right in figure 4-1. Good flow distribution and satisfactory turbulence levels occur when the flow rate through the baffle averages about 45 liters per second per square meter. The head loss for this flow rate is about 50 millimeters of water. Good flow distribution, but higher turbulence and water-surface roughness, can be obtained with a flow of 90 liters per second per square meter. A head loss of about 100 millimeters should be allowed for the 90-L/(m$^2 \cdot$s) flow rate. For large spans, the 100-mm-thick baffle may need intermediate stiffeners or the thickness of the cross section can be increased to 150 millimeters or more. Parallel sheets of corrugated metal or wire screen baffles containing aluminum shavings or plastic fibers may also be used for uniformly distributing the flow in a model reservoir.

Methods of constructing models of overflow spillways are many. The use of metal templates, hand- or machine-formed to the spillway profile, works well on straight sections (fig. 4-5). The templates are bolted together in a metal-and-wood frame; then cement-sand mortar is placed in the spaces between the templates, either while the spillway is on a workbench or after it is assembled on the model. A final smooth coating of grout is applied with a metal screed having a shape identical to the profile of the crest, or the smooth surface can be obtained by troweling on and then sanding the finish coat of grout. Instead of using cement-sand mortar to obtain the final

Upstream view.
P801-D-74601

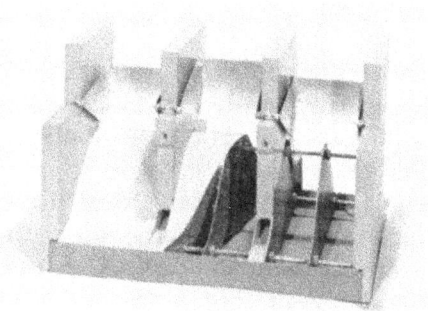

Downstream view.
P801-D-74602

Figure 4.5.–Model of gated spillway.

spillway profile, thin sheet metal can be formed to fit over the profile templates and then soldered in place.

In the course of preparing the template assembly, piezometer tubes are soldered or otherwise fastened along the profiles of selected templates. If the spillway crest is to be formed with mortar, the pressure taps are closed with lightly fitting plugs before applying the mortar and grout. Upon completion of the model, each tap should be carefully reopened and then finished flush with the surface of the spillway. For spillways formed of sheet metal, the piezometer tubes are inserted through holes drilled in the sheet metal, soldered in place, and finished flush with the surface. The finishing of piezometers in models should be done meticulously to prevent measurement errors that would result from improper installation. Sizes of tubing for connecting piezometers to manometers or pressure transducers should be selected from

68

1-mm-minimum inside diameter to possibly 3-millimeter maximum, depending on the size of the model.

Complicated curves for tunnel spillway crests, bends, and transitions can be made from plastic which has been heated and formed over a prepared wooden pattern. Acrylic plastic sheets, available in various thicknesses, are well suited for this type of construction. The fabrication techniques used may be developed by the shop patternmaker or an associate, provided he has a knowledge of proper temperatures for heating, other characteristics of the plastics, machining and bonding methods, and moisture sensitivity, all of which can normally be obtained from the manufacturer. Piezometer tips (fig. 4-6) are machined from the same material as used in forming the plastic model. The tip is cemented to the outside of the model at the desired point and a hole is drilled through the model wall using the tip as a drill guide. Burrs should be carefully removed from inside the drilled hole. Flexible or rigid tubing is used to connect the piezometer tip to a manometer or pressure transducer.

Plastic materials differ widely in physical properties. Thus, the engineer and plastics craftsman must cooperatively work out details of construction to obtain satisfactory accuracy of shape in the model.

Rigid foam plastics are construction materials adaptable to architectural or operating hydraulic models. Tools used in forming model sections from such materials are available in a machine or carpentry shop (fig. 4-7). Waterproof coatings of paint or epoxy are used to smooth the crest and provide a satisfactory surface roughness. A crest of this type of plastic is light and easy to handle, but, since it is relatively soft, care must be taken to prevent damage.

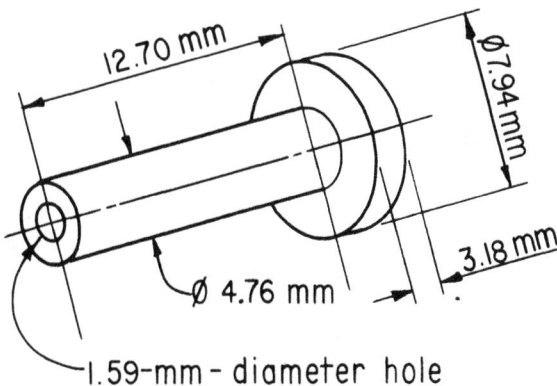

Figure 4-6.–Piezometer tip for 6.35-mm-outside diameter flexible tubing. 103-D-1712.

Figure 4-7.–Curved spillway crest of rigid foam being formed on drill press using a routing bit and template. P801-D-79226

Semirigid plastic or metal may be used for the piezometer tubes, which must be carefully installed to provide a smooth flush surface at the intersection of the tubes and the soft material of the crest.

Wood exposed to water is not recommended for the more permanent portions of models because of shrinking and swelling. Wood with a waterproof coating is satisfactory for constructing piers and training walls of spillways because these parts are usually subject to modifications.

Models of gates and valves may be simple or extremely complex to construct. A gate model may be simply a thin sheet of metal used to produce a jet of water resembling that in the prototype, or it may have to be reproduced in great detail to obtain proper scaling of loads caused by water submerging both upstream and downstream parts of the gate. An example of such a model is shown in figure 4-8. Here, one gate of a multiple-gate structure was modeled to determine lift forces and vibration characteristics. A strain gage balance supporting and restraining the gate in an upstream direction was designed coincident with the model. This model of one bay of the structure provided enough information to assure that the design was satisfactory.

70

Figure 4-8.–Structurally detailed gate model with strain gage balance for study of lift and vibration forces. P801-D-74585

A machine shop is indispensable to a well-equipped hydraulic laboratory. Gates and valves used in detailed studies must be machined with precision (fig. 4-9). Satisfactory model investigations require careful consideration of location of piezometers, smoothness of surfaces, and ease of operation. Selection of materials for model gates and valves is based on the required useful life of the model. Brass is normally used because it casts and machines well. Aluminum alloys may be used, but corrosion causes deterioration and binding of moving parts in a short period of time. Commercial gears and other standard parts should be considered in the construction of gate and valve models. Attention to detail in the model design by the hydraulic engineer will ease the work of the machinist in producing a gate or valve model of the required accuracy.

4-4. Operation of Model.–The operating program should be carefully planned to properly evaluate the worth of the design under study. In general, the evaluation includes proving, by qualitative and quantitative tests, that the design meets the operation requirements. For an undistorted model, the operating program can be divided into two phases–adjustment and testing. In the case of a distorted, movable-bed model, a prototype verification must also be considered.

Inlet. P801-D-79225

Outlet.
P801-D-74603

Figure 4-9.–Precision-machined, 150-millimeter (6-in) model of a hollow-jet valve.

The adjustment phase includes preliminary trials to reveal model defects and inadequacies. This important phase should not be hurried; time should be taken to make certain that the model performs as intended and that the instrumentation is satisfactory. The need for partial redesign, revision, or shifting of measuring instruments is often indicated by these trial tests. During the adjustment phase, the investigator becomes acquainted with the peculiarities of the model and becomes adept in its operation.

When a model has a movable bed, the adjustment phase involves verification of the available recorded action of the prototype. This action must be duplicated in the model; otherwise, reliable quantitative results cannot be obtained. Because such tests are often the only basis for similitude, verification of a recorded prototype action is an extremely important part of the operation of movable-bed models.

Testing should include a systematic examination of each feature of a proposed design for operation improvement, possible reduction in cost of construction, and reduction in maintenance costs. The investigator must exercise patience, imagination, and ingenuity, and be capable of interpreting the model results correctly. He should work in close cooperation with the design engineers, continually informing them of progress and consulting with them concerning future testing. Data should be analyzed concurrently with the testing to prevent accumulation of unnecessary information. As the study progresses, functional relationships among the different variables should be examined to aid in detection of measurement errors.

In addition to the regular testing, the investigator should obtain information to aid in generalizing the results of studies made on other structures of the same type. Results of studies on many models are usually required in assembling general design information. It is a waste of time and money to construct models and not obtain general as well as specific information. For example, the following general information is desired on spillway and outlet works designs:

- Calibration curves for spillway crests,
- Complete calibration curves for gates,
- Water surface profiles through gate sections of spillways and other open-channel structures,
- Information on the spreading of jets in flat chutes,
- Stilling pool action for various types of sloping aprons and baffle-pier arrangements, and
- Effects of negative pressures on the flow stability and discharge capacity of overfall crests.

Some of these data are obtained in routine testing, but care should be taken to obtain the remaining general information when possible. Such information can be valuable for verification studies by actual field measurements.

Because the end product of any model study is the report which transmits the findings and recommendations, the investigator must maintain a complete and accurate set of notes on measurements and observations, and keep a diary, since dates may be of special significance in the future. Negative as well as positive results should be recorded. A complete photographic record of all important tests is indispensable and often eliminates the necessity of repeating tests. Video tape records of portions of the study for comparison of progressive stages serve to recall the effect of model changes. The tapes may be reviewed while observing model operation of proposed changes. The importance of presenting a clear, concise, well-organized, and well-illustrated report of the investigation cannot be overemphasized.

CHAPTER 5

Free-Surface Flow

A. INTRODUCTION

5-1. General.–Hydraulic laboratory techniques applied to open-channel models are discussed in this chapter. Types of open-channel structures and basic requirements for models that will provide accurate test data are described to give the investigator a background for analyzing the particular design under study. A knowledge of the designer's viewpoint and the assumptions made in designing the hydraulic structure are also essential to the laboratory investigator so that practical solutions can be obtained from a hydraulic model.

B. SPILLWAYS

5-2. General.–Spillways are constructed to various profiles. The most common spillway in concrete dams is the free fall, or overfall type, having a curved crest designed for a particular head (fig. 5-1). Open-chute spillways (fig. 5-2) having an overflow crest are often used on earth dams, and tunnel spillways are used in narrow canyons or at sites where an open channel would be subjected to snow or rock slides. The spillway crests may be free overflow without controls, or the flow may be controlled by radial, slide, or wheel-mounted gates.

Usually, spillway models are built geometrically similar to their prototypes; gravity forces predominate and dynamic similitude is closely approximated according to the Froude parameter. Except in models having long chute spillways or tunnels where viscous effects distort the velocities or depths of flow, viscosity and surface tension can ordinarily be neglected if a proper scale is chosen for the model. For example, model spillway gates should be at least 150 millimeters in width for crest pressure studies or for determination of discharge

Figure 5-1.–Overfall spillway and plunge-type stilling basin. P801-D-79224

coefficients. If the prototype has several gates and this dimension cannot be attained because of space limitations in the laboratory, a partial model of the crest with only one or two of the gates represented will give satisfactory data.

For studying erosion and energy dissipation downstream from a chute or tunnel spillway, the velocity of flow leaving the model tunnel or chute must approximate the scaled prototype velocity. Corrections for the proportionally higher friction losses of the model may be made by increasing the slope, by decreasing the length of a tunnel or chute, or by combination of both. To minimize distortion of the flow pattern in a spillway tunnel having both inclined and horizontal sections, the correct slope may be maintained and the correct velocity obtained by shortening the horizontal portion of the tunnel. The increased flow velocity may also be obtained by increasing the model reservoir head.

Theoretical velocities and depths in the model and prototype spillway or tunnel may be computed using a reliable resistance relationship such as the Manning equation:

$$V = \frac{1}{n} R^{2/3} S^{1/2}$$

Figure 5-2.–Open-chute spillway model. POA-27D-67599NA

In most cases, the following values of roughness may be assumed: $n = 0.014$ for the prototype, and 0.010 for the model. Velocities are computed to the exit of the prototype tunnel or spillway. A similar computation is made for the model. Any major differences are compensated for by a physical adjustment in the model.

Table 5-1 shows how to determine a tunnel length correction from velocity computations. A tunnel spillway with a morning glory inlet had been designed so that the throat at elevation 1073.810 meters was the control. The tunnel was to flow full, with a negative head of 1.95 meters for the maximum discharge of 1500 cubic meters per second. Computations were started at the control, using a roughness coefficient n for the prototype of 0.014. Loss in the bend immediately below the crest was estimated to be approximately 2.4 meters of water; in a free-flow bend upstream from the outlet, loss was estimated to be negligible. Velocity at the outlet portal of the spillway was computed to be 44.60 meters per second.

The velocity at the outlet portal of a complete 1:36 scale model of the spillway tunnel was computed in the same manner as for the

Table 5-1.—*Prototype spillway velocity computations. 103-D-1745.*

SECTION THRU SPILLWAY

DATA
- Max WS EL 1086.58
- Max Q 1500 m³/s
- Designed neg head at throat 1.95 m
- Computed upper bend loss 2.44 m
- Head loss at lower bend assumed to be zero
- Head losses to throat EL 1073.81 assumed to be zero
- Roughness coefficient "n" estimated to be 0.014

FORMULAS

$$V = \frac{Q}{A}, \quad h_V = \frac{V^2}{2g}, \quad S = \frac{n^2 V^2}{R^{4/3}}, \quad h_f = SL$$

1 ℄ STA	2 ℄ EL	3 Invert EL	4 D	5 dn/D	6 dn	7 dn Cos θ	8 A	9 V	10 hv	11 R	12 S	13 Avg S	14 L	15 hf	16 Σhf	17 dn Cos θ +hv+Σhf	18 EGL + losses EL
1+54.534	1073.810		10.604	1.00	—		88.31	16.99	14.718	2.652	0.0154	—					
1+60.096	1057.613		10.604	0.72	7.634	4.907	68.07	22.04	24.767	3.164	.0205	0.0180	17.511	0.315	2.752	32.426	1086.63
1+89.171	1022.964		9.559	.59	5.640	3.625	44.05	34.05	59.113	2.630	.0627	.0416	45.232	1.882	4.634	67.372	1087.26
2+18.246	988.314		8.513	.60	5.108	3.283	35.66	42.06	90.196	2.362	.1105	.0866	45.232	3.917	8.551	102.030	1087.61
2+47.318	953.667		7.468	.67	5.004	3.217	31.20	48.08	117.863	2.179	.1606	.1356	45.232	6.133	14.684	135.764	1087.03
2+66.115	—	936.803	7.468	.65	4.854	4.854	30.14	49.77	126.295	2.152	.1750	.1678	23.927	4.015	18.699	149.848	1086.65
3+04.142	—	936.724	7.468	.60	5.078	5.078	31.72	47.29	114.022	2.192	.1542	.1646	38.039	6.261	24.960	144.060	1080.78 *
				.67	5.004	5.004	31.20	48.08	117.863	2.179	.1606	.1678		6.383	25.082	147.949	1084.67
3+42.165	—	936.660	7.468	.69	5.153	5.153	32.23	46.54	110.434	2.203	.1484	.1545	38.039	5.877	30.959	146.546	1083.21 *
				.68	5.078	5.078	31.72	47.29	114.022	2.192	.1542	.1574		5.987	31.069	150.169	1086.83
3+92.762	—	936.565	9.449	.43	4.063	4.063	32.74	45.82	107.043	2.225	.1419	.1480	50.597	7.488	38.557	149.663	1086.23
4+27.357	—	936.498	9.449	.44	4.158	4.158	33.63	44.60	101.419	2.259	.1317	.1368	34.595	4.733	43.290	148.867	1085.37

Column 5 assumed.
Column 17 equals depth plus velocity head plus losses.
Column 18 equals invert elevation plus Column 17.

* The assumed value of dn in Column 5 is not close enough to true value.

All dimensions are in meters unless shown otherwise.

prototype, making the same assumptions except the roughness coefficient n was assumed to be 0.010 for the plastic model. Computed model velocity at the outlet portal was 6.724 meters per second, equivalent to 40.34 meters per second in the prototype according to Froude's law.

To increase the velocity at the outlet portal in the model to more nearly represent the prototype, 1.52 meters of a 2.11-meter length of uniform circular section downstream from the lower bend was eliminated. Omission of this length in the model was assumed to not affect appreciably the flow patterns through the lower bend and a horseshoe transition. The velocity of the model was then computed to be 7.066 meters per second, which corresponds to 42.40 meters per second in the prototype (see table 5-2). Eliminating other portions of the tunnel was not desirable. Thus, the value of 42.40 meters per second was accepted as being sufficiently close to the computed prototype value of 44.60 meters

Table 5-2.–*Model spillway velocity computations. 103-D-1746.*

1:36 MODEL SPILLWAY VELOCITY COMPUTATIONS
(76.05 meters of 7500-mm-diameter conduit downstream from lower bend is shortened to 0.59 m in the model)

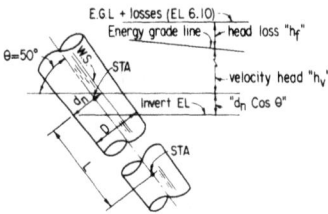

DATA

• Prototype max. W.S. EL 1086.58 = model EL 6.10.
• Prototype Q of 1500 m³/s = model Q of 193 L/s.
• Designed neg. head at throat of 1.95 m proto. = 0.055 m model.
• Computed upper bend head loss of 2.44 m proto. = 0.068 m model.
• Head loss at lower bend assumed to be zero.
• Head losses to throat (model EL. 5.74 m) assumed to be zero.
• Roughness coefficient "n" estimated to be 0.010.

FORMULAS

$$V = \frac{Q}{A}, \quad h_v = \frac{V^2}{2g}, \quad S = \frac{n^2 V^2}{R^{4/3}}, \quad h_f = SL$$

1	2	3	4	5	6	7	8	9	10	11	12	13	14	15	16	17	18
₵ STA	₵ EL	Invert EL	D	$\frac{d_n}{D}$	d_n	$d_n \cos\theta$	A	V	h_v	R	S	Avg. S	L	h_f	Σh_f	$d_n \cos\theta + h_v + \Sigma h_f$	E.G.L + losses EL
1+54.534	5.742		0.294	1.00			0.0681	2.832	0.4089	0.0738	0.0260					0.354	6.096
1+60.096	5.291	5.197	0.294	0.72	0.212	0.1363	.0525	3.676	0.6890	.0878	.0346	0.0303	0.4877	0.0146	0.0823	.908	6.105
1+89.171		4.243	0.265	.60	.159	.1022	.0347	5.567	1.5801	.0738	.1004	.0675	1.2558	.0848	.1671	1.849	6.092
2+18.246		3.290	0.237	.61	.144	.0926	.0281	6.879	2.4127	.0661	.1771	.1388	1.2558	.1743	.3414	2.847	6.137
2+47.318		2.337	0.208	.69	.143	.0919	.0249	7.751	3.0631	.0613	.2487	.2129	1.2558	.2674	.6088	3.764	6.101
2+66.115		1.935	0.208	.67	.139	.139	.0241	8.022	3.2811	.0607	.2700	.2594	0.6645	.1724	.7812	4.201	6.136
2+82.146		1.934	0.208	.69	.143	.143	.0249	7.749	3.0615	.0613	.2500	.2600	.5852	.1522	.9334	4.138	6.072
3+37.743		1.932	0.262	.44	.116	.116	.0259	7.438	2.8207	.0628	.2222	.2361	1.4051	.3317	1.2651	4.202	6.134
3+72.338		1.930	0.262	.46	.121	.121	.0273	7.066	2.5456	.0643	.1936	.2079	0.9601	.1996	1.4647	4.131	6.061

Column 5 assumed.   All dimensions are in meters unless shown otherwise.
Column 17 equals depth plus velocity head plus losses.
Column 18 equals invert elevation plus Column 17.

per second to be satisfactory, and the spillway tunnel was modeled in this manner.

Similar computations are made for long chute spillways. Slight adjustment to the slope of the chute provides the proper exit velocity for energy dissipation in the model. For some investigations, a shortening or lengthening of the chute may not be desirable because of the positions of reflected waves.

5-3. Overfall and Overflow Spillways.-(a) *Spillway Capacity and Pressures.*-The results of model tests are an aid in the design of overflow structures to ensure adequate and safe operation of the prototypes. Spillway capacity measurements must be accurate. A small error in the model measurement may be equivalent to a considerable quantity of water in the prototype because of the large scale ratio for discharge conversion.

The effect of submergence on the capacity should be investigated if the crest is to be submerged during spillway operation [16]. (When the water surface elevation in the channel below a spillway is at or above the elevation of the crest, the crest is said to be submerged.) The depth downstream from the crest may not affect the spillway capacity until the ratio of downstream to upstream depths, referred to the crest elevation, reaches 0.7 to 0.8. Even in studies where the crest is not submerged as judged by visual inspection, the effect on the discharge coefficient should be determined if the head is approximately equal to the vertical distance from crest to apron. The proper length of spillway crest necessary to obtain the desired prototype discharge for the expected operating conditions may be calculated using the discharge coefficients determined from the model. The recommended spillway should be studied to assure that unpredictable end contractions, unusual approach conditions, or other factors do not cause a reduction in capacity. Additionally, the recommended spillway should be studied to provide accurate rating curves for operating the prototype structure.

Before concluding the studies, the spillway surface profile should be studied for the existence of subatmospheric pressures. An example of a high-coefficient crest and one prone to subatmospheric pressures is the overfall shape shown on figure 5-3. For most spillways, the pressures on the face should be atmospheric or only slightly subatmospheric if the structure is to operate frequently near its maximum capacity.

Very little is known about prototype spillways operating at subatmospheric pressures. A partial vacuum may be relieved at intervals by air breaking through the nappe surface or between the

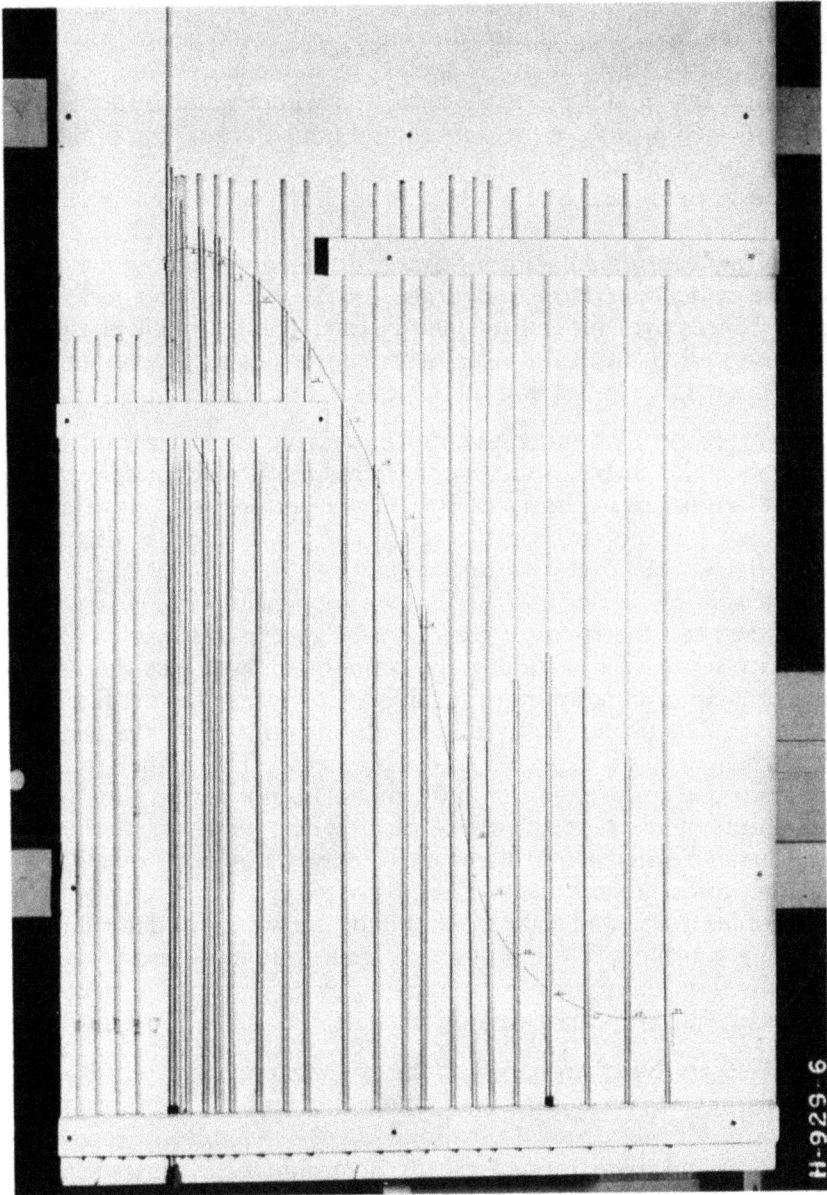

Figure 5-3.–Pressures on a 5-percent vacuum overfall crest spillway. P801-D-79223

nappe and a side boundary. Such relief would probably be periodic and would cause fluctuating pressures on parts of the structure. Fluctuating pressures are undesirable and may cause dangerous vibrations to be transmitted to gates, valves, or other mechanical equipment. As a general rule, pressures on a prototype spillway face should not be lower than 4.5 meters of water below atmospheric pressure. For model pressures measured with a water manometer, the indicated prototype pressure should not be lower than 3 meters below atmospheric.

The appearance of the flow in models will often indicate difficulties not apparent in data obtained from the discharge and pressure measurements. Poor flow conditions approaching the crest will be evident in contractions at the ends of the crest and in surface disturbances over the entire spillway face. The hydraulic engineer becomes adept in using visual observations as a complement to instrumentation in judging the adequacy of a design.

(b) *Flow Characteristics.*–Excessive end contractions, improper pier shapes, and very sharp radii in flow boundaries will result in rough water surfaces. For proper performance, flow in the approach channel and on the spillway face should be relatively uniform without excessive turbulence.

The spillway shown in figure 5-2 is a good example of satisfactory flow between the spillway crest and the energy dissipator. Water velocity across each section of the chute or spillway face should be uniform. Interpretation must include an allowance for the fact that small models do not fully indicate the magnitude of the surface disturbances to be expected in the prototype. The higher relative viscosity and surface tension in the model do not permit scaling the air entrainment. A small surface disturbance, which in the model appears to be attached to the main sheet of water may be detached in the prototype and result in excessive spray.

Profiles of the water surface should be measured for determining training wall height, for positioning gate trunnions, and to assure that the flow will pass gate counterweights or parts of the structure not reproduced in the hydraulic model.

(c) *Gate Operation.*–When the spillway crest shape and other details have been determined, the gates may be calibrated for discharge through partial openings. Curves or tables of reservoir elevation and discharge are usually determined for representative or probable gate openings. Then, computed gate discharge coefficients can be conveniently used to prepare curves or tables for the prototype spillway flow.

In model tests, consideration should be given to the spillway performance while the gates are being opened. A rapid opening of the gates can result in dangerous flow conditions in the energy dissipator. To prevent costly mistakes in prototype operations, a gate operating schedule should be developed using the model test results. The proper sequence for opening and closing the gates, the limiting increment, and the uniformity of gate openings should be carefully determined to minimize asymmetrical flow and eddies. Eddies can carry abrasive material into the stilling basin and cause serious concrete erosion. The maximum rate of change of gate opening should be determined for satisfactory operation of the energy dissipator to minimize riverbed erosion. Experiments at various discharges should be performed to determine at what tailwater elevation the water will be swept from the energy dissipator.

If the spillway under test has flow over the gates, as in the case of drum or hinged-leaf gates, the problem of ventilation should be investigated since the action of the flowing water tends to evacuate air from the space beneath the nappe. If a continuous supply of air is prevented from entering this space because of the nappe sealing against the spillway piers, a vent should be provided to prevent pressure reduction, which can cause instability of flow and an added load on the gate. Any fluctuation in pressure can result in vibration or oscillation which, under unfavorable conditions, might result in failure of the gate.

In addition to the factors discussed, there are usually other special problems associated with each particular structure. Determination of the size and type of gate may be necessary. Pressures on gate surfaces, unbalanced pressure on spillway piers or training walls, proper curvature of training walls, and special hoods or deflectors on the spillway face may be items for investigation.

5-4. **Tunnel Spillways.**–There are two predominant types of tunnel spillways: those having inclined shafts (fig. 5-4) and those with vertical shafts (fig. 5-5). The collecting structure at the entrance to an inclined-shaft spillway may be any of the following types: side channel, semicircular overflow weir, straight overflow crest, or morning glory crest (fig. 5-5). The vertical-shaft spillway of large capacity generally has a morning glory crest. The collecting structures may or may not have gates. In all of these designs, a transition is required between the collecting structure and the tunnel. The inclined and vertical tunnel profiles are similar in design and usually include a converted part of the diversion tunnel. Structures having inclined shafts usually operate with open-channel flow to simplify aeration, flow, and pressure problems. Structures with vertical shafts are designed to operate either with free flow, submerged flow (vertical shaft flowing partially full), or both.

A. Entrance and crest with raised gates. (Paper confetti on water surface for detecting flow patterns and determining velocities.) P622-D-45135NA

B. Inclined tunnel. P622-D-45163NA

Figure 5-4.–Model of overflow spillway and tunnel.

A. Free-flow crest at partial discharge. (Paper confetti on water surface for detecting flow patterns and determining velocities.) P805-D-46309NA

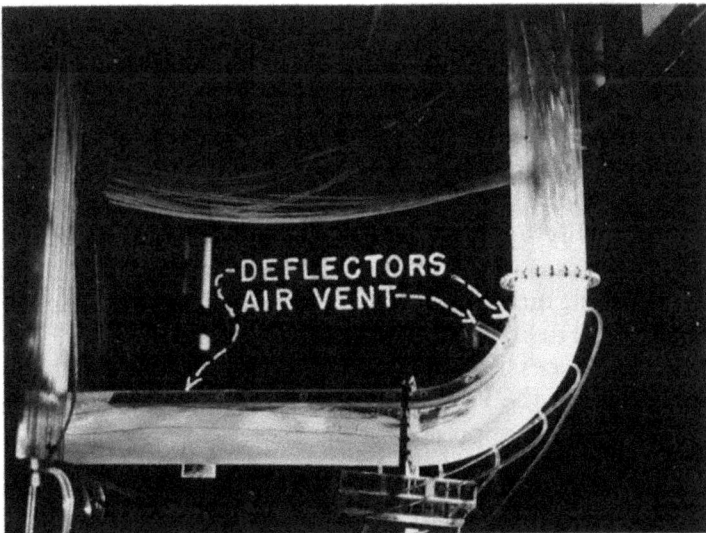

B. Tunnel bend. PX-D-36251

Figure 5-5.–Model of morning glory spillway and tunnel.

(a) *Pressure and Flow Conditions.*–Pressures within the spillway entrance shaft and the discharge tunnel are of major concern. Severe subatmospheric pressures can produce undesirable vibration, noise, and possibly cavitation erosion of the spillway surfaces. Piezometer openings in the model should be located with care and in generous numbers, especially on the morning glory spillway stuctures. Extensive studies of pressures on the entrance and in the throat of the vertical portion of the shaft should be made. Subatmospheric pressures may require ventilation and the model serves to determine the size and location of required air vents.

Smooth flow should occur throughout the structure, whether it operates as a closed conduit or as an open channel. So far as possible, the approaching flow should be normal to the crest of the collecting structure, and its depth over the crest should be uniform. Flow within the vertical or inclined shaft should have a minimum of turbulence to minimize air entrainment. A converging shaft or deflectors at bends in the shaft assist in maintaining positive or near positive pressures on the walls of the shaft. The flow in the bend connecting the vertical or inclined shaft with a horizontal free-flow tunnel should neither tend to seal the tunnel nor cause the flow in the tunnel to exceed design depth. Sharp horizontal bends, which tend to cause the tunnel to intermittently flow full, should be avoided. Such a condition may cause fluctuating pressures and create unstable flow, accompanied by intense surging, noise, and air pumping. Study of pressure distribution and fluctuations in these structural components must be thorough to assure satisfactory prototype operation.

(b) *Entrance and Exit Considerations.*–One of the most interesting, but difficult, areas of study in connection with submerged morning glory spillways is the elimination of vortices, which invariably form within the intake structure. Vortices are undesirable because they cause reduction in flow capacity, excessive air entrainment, turbulence, and, possibly, vibration. Many devices have been tried to prevent vortices from forming, but none has been universally successful. Methods commonly used with success include piers on the crest, flow guides within the entrance, and control of the flow approaching the crest.

The tunnel exit should be designed for free flow because unstable flow conditions can occur if the discharge alternately seals and then flows freely, causing fluctuating pressures and objectionable vibrations. Entrained air compressed near a submerged tunnel outlet may discharge with explosive force from the portal. The portal should be placed sufficiently high to prevent a hydraulic jump from forming in the tunnel for all but minimum discharges. If the spillway must be operated with the outlet near submergence for large flows, pressure and velocity data from a large model should be used to carefully evaluate the flow action.

C. STILLING BASINS

5-5. General.–Stilling basins, or energy dissipators, are constructed in conjunction with spillways, outlet works, or other structures that discharge high-energy water. These basins protect the main structure and the downstream channel from damage by high-velocity water. There are several types of stilling basins, each being appropriate under a definite set of conditions. Laboratory investigators should understand the various types to enable proper judgment to be made of the performance of any one.

5-6. Types of Stilling Basins.–The most common type of stilling basin is the hydraulic jump basin with horizontal apron, where the energy is both dissipated and distributed before the water is released into the downstream channel. This basin, shown in figure 5-6A, is most suitable when the depth D_2 computed from the formula:

$$D_2 = -\frac{D_1}{2} + \sqrt{\frac{D_1{}^2}{4} + \frac{2V_1{}^2 D_1}{g}}$$

coincides with or closely approximates the tailwater depth for the river for all discharges; the near coincidence of the D_2 curve (curve A, fig. 5-7) and the tailwater curve means a good hydraulic jump will form. The length of apron necessary to develop the jump fully ranges from 4 to 6 D_2, depending on the Froude number of the incoming flow. This length can be reduced to 2 to 3 D_2 by the addition of chute blocks and baffle piers to the apron as shown on figure 5-6B. Such stilling basin appurtenances also provide a considerable factor of safety against jump sweepout if the tailwater elevation should be lowered by future degradation of the riverbed downstream from the dam.

Should the computed D_2 curve differ from the tailwater curve of the river, as in curve B of figure 5-7, study of the proposed stilling basin is necessary for the range of discharges having a deficiency of tailwater. The performance of a horizontal apron may be

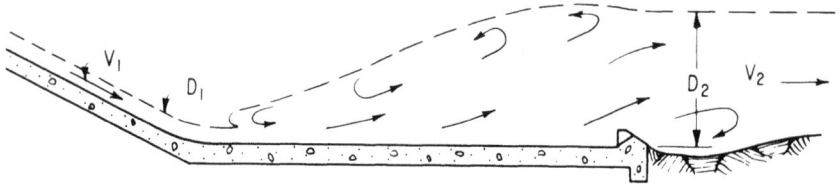

A - HORIZONTAL APRON

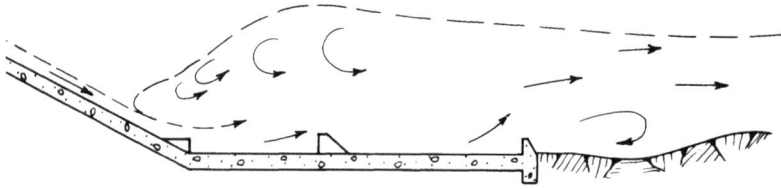

B - HORIZONTAL APRON WITH
CHUTE BLOCKS AND BAFFLE PIERS

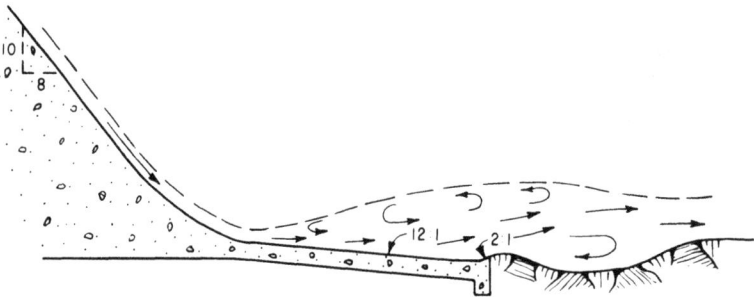

C - SLOPING APRON

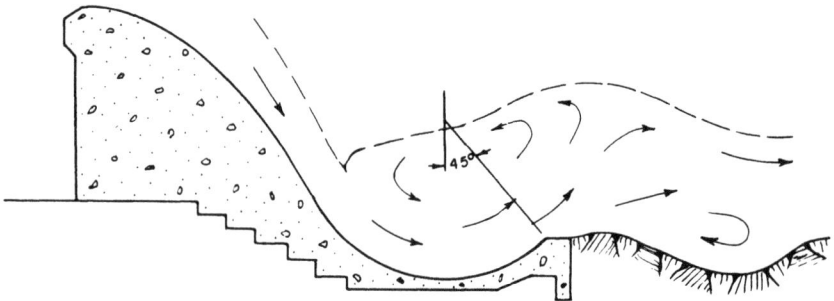

D - ROLLER BUCKET

Figure 5-6.–Stilling basins for spillways and outlet works. 103-D-1713.

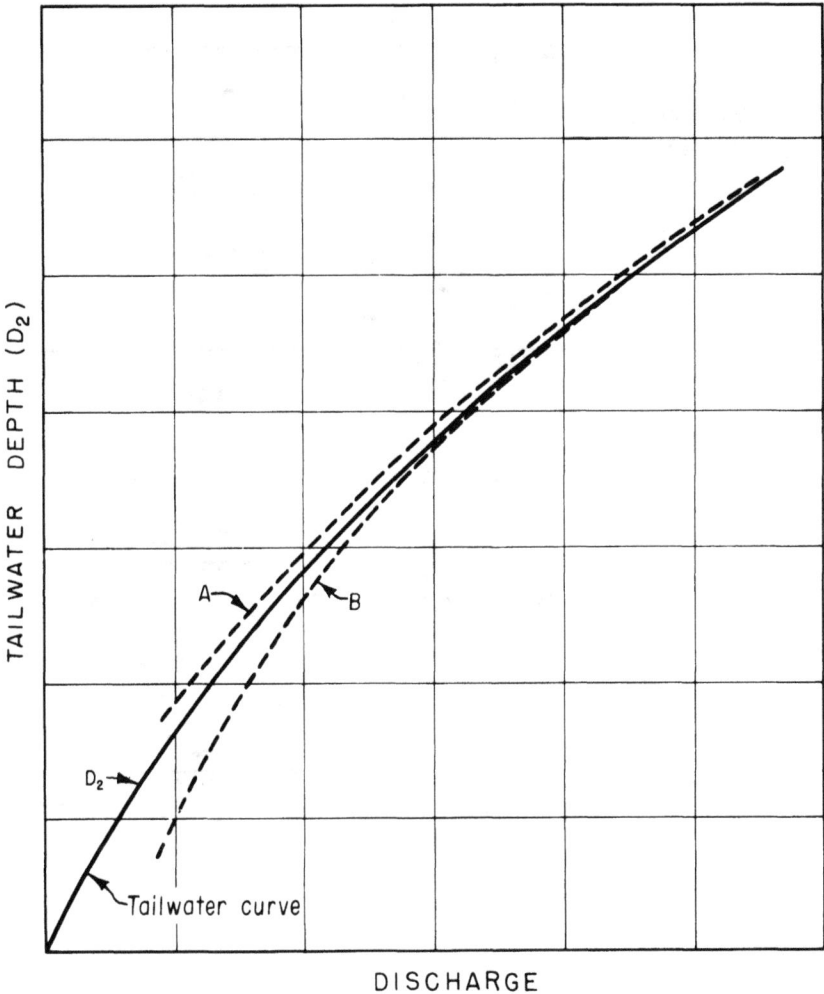

Figure 5-7.–Relationship of the computed D_2 curve for a horizontal apron stilling basin to the actual tailwater depth in the exit channel. 103-D-1714.

predicted with a good degree of accuracy by determining the Froude number of the incoming flow and selecting from figure 5-8 the type of jump action that will occur.

Figure 5-6C shows a sloping apron and figure 5-9 shows some of the design charts and drawings useful in determining the dimensions of the basin [17]. A sloping apron requires greater depth and length to contain the jump than does a horizontal apron.

Should the computed D_2 curve fall definitely below the tailwater curve, the stilling basins described above are not applicable. Where

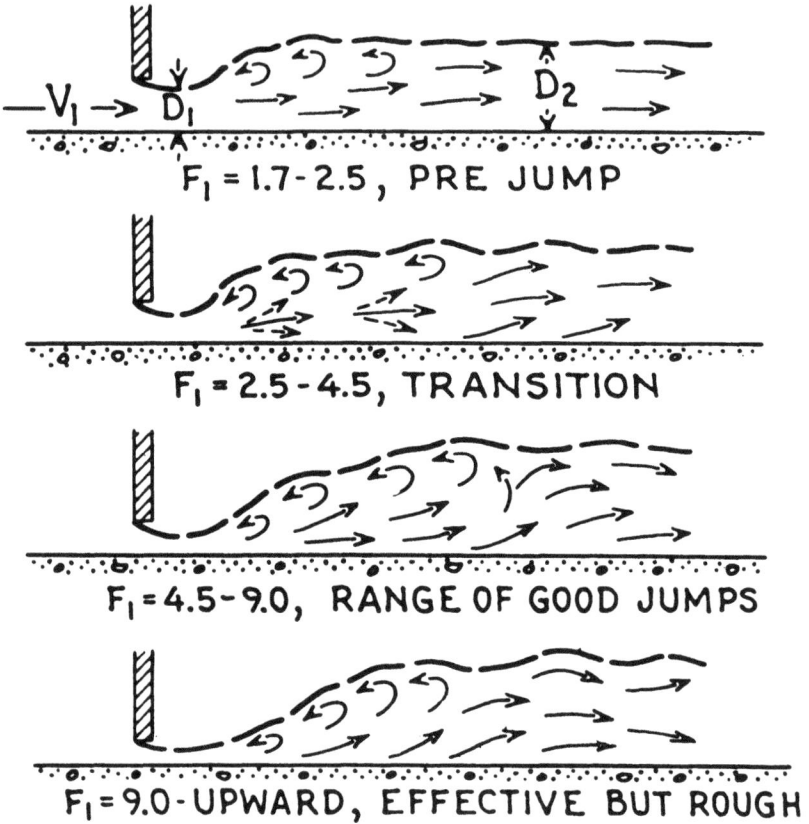

$F = $ Froude number

Figure 5-8.–Hydraulic jump forms. 103-D-1715.

the tailwater depth is excessive, a curved apron or bucket dissipator may be feasible. The roller bucket (fig. 5-6D) is effective for tailwater depths which greatly exceed those of the conventional apron type of basin. This type of bucket was used on the Grand Coulee Dam spillway and performs best when deeply submerged. A disadvantage is that a violent ground roller develops downstream. Unsymmetrical flow resulting from this ground roller can deposit large boulders on the apron and cause erosive damage to concrete. Symmetrical flow, on the other hand, will prevent this abrasive material from entering the apron. A model study of pressure, velocity distribution, and erosion tendencies is advisable when this type of stilling device is under consideration.

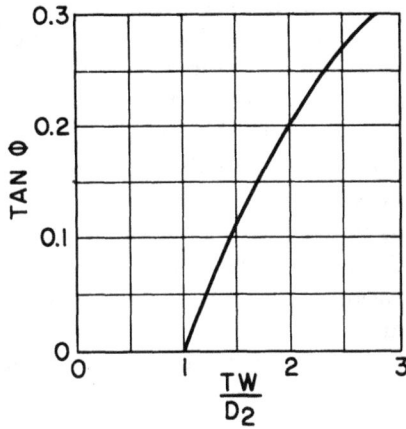

TAILWATER
DEPTH RELATED TO
CONJUGATE DEPTH

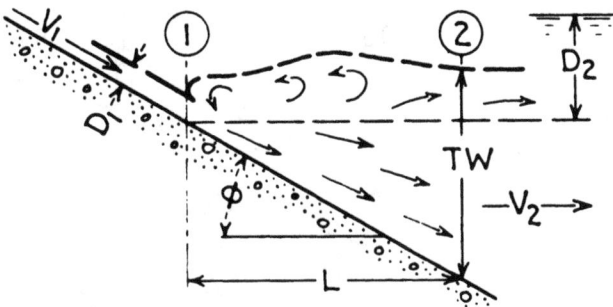

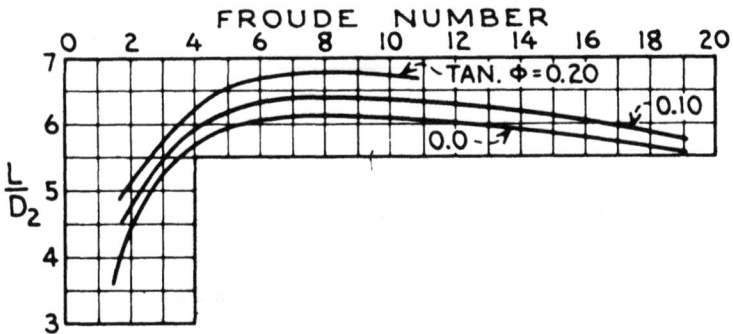

LENGTH OF JUMP

Figure 5-9.–Design curves for sloping aprons. 103-D-1716.

The slotted bucket (fig. 5-10) is a modification of the roller bucket and differs from the solid type because of the slots and the short downstream apron. In comparison with the solid bucket, the hydraulic action of the slotted bucket results in less large-scale turbulence and waves and a less violent ground roller. Water flowing through the slots spreads to counteract a portion of the ground roller. The slotted bucket operates well over a wide range of tailwater depths but requires a tailwater depth somewhat greater than that required for a conventional horizontal-apron stilling basin. The usual dimensions of the energy dissipator are given in terms of the bucket radius [17].

A third type of spillway energy dissipator, sometimes called a flip bucket (see fig. 5-11), discharges the flow through the air into the water downstream from the dam and may be used for practically any tailwater condition. This bucket may be near or far above the

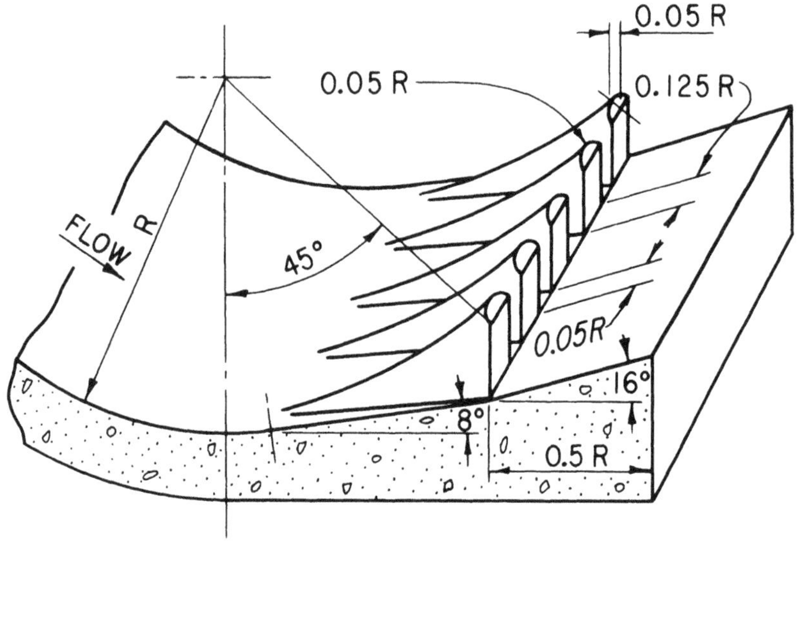
0.05 R
0.05 R
0.125 R
FLOW
R
45°
0.05R
16°
8°
0.5 R
Notes: For use on spillways, drops, chutes etc. where crest flow is not submerged. V_1 and F_1 have no limits.
Data may also be used to design a solid-type roller bucket.

Figure 5-10.–Slotted-bucket energy dissipator. 103-D-1717.

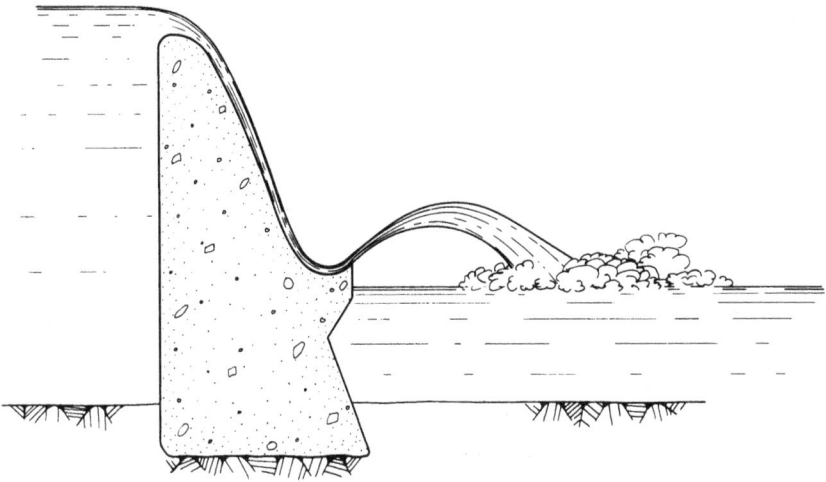

Figure 5-11.–Flip-bucket energy dissipator. 103-D-1718.

tailwater surface. The energy of the water is dissipated in the stream or a preformed plunge pool downstream from the structure. The impact of the flow produces rough water surfaces and spray that may be undesirable for nearby structures such as powerplants and switchyards. A model study is often needed to develop the design of a flip bucket and plunge-type energy dissipator such as shown in figure 5-11.

5-7. **Visual Appraisal of Action.**–Stilling pool appurtenances such as chute blocks, floor blocks, baffle piers, end sills, or deflectors may be used to improve energy dissipation and reduce the basin cost. However, the safety of the structure should not depend primarily on these relatively small accessories, which may be damaged or lost during periods of continued operation.

A properly operating stilling basin has a mildly undulating water surface, no significant waves, and no local scouring velocities. Uniform velocity distribution at the downstream end of the basin and a minimum of channel erosion is essential to good operation. The relative effectiveness of different basin designs may be determined by comparing measured exit velocities, wave heights, apron and wall pressures, and erosion of the downstream riverbed. Observation of the general appearance of the operation, preferably recorded as a series of still photographs or as movies or video tapes, will facilitate evaluation of the effectiveness of a design. In making studies of a series of stilling basin designs, the procedure to be followed will vary from one study to another. However, by following a predetermined plan of testing, unnecessary work can be avoided and a maximum of information can be obtained.

Evaluation of various designs by the general appearance of hydraulic action is advantageous in the study of preliminary or experimental designs. By watching the performance for a short time, an experienced investigator can usually predict whether a proposed design shows promise or whether the plan should be abandoned. Often, only a visual appraisal of the performance of a particular design is needed. If it appears that with minor changes the design can be made workable, other tests may be made to corroborate the investigator's judgment. The presence of extreme turbulence, waves, high velocities, severe eddies, or pulsating flow within or beyond the limits of the structure indicates need for further refinement of design or complete abandonment of the scheme.

If preliminary observations show that further testing is required, the basin should be studied for velocity and pressure distribution at flows covering the range of discharges and headwater elevations to be expected in the prototype. For selected discharge and headwater conditions, the permissible range of tailwater elevations should be determined by lowering the tailwater elevation until the water sweeps out into the downstream channel. Basins that do not operate over a fairly wide range of tailwater elevations are not ordinarily practicable. Unpredicted lowering of the tailwater from erosion of the streambed over a long period of time or rapid lowering from a reduction in powerplant discharge could result in an unsafe energy dissipator if the model studies failed to include low tailwater conditions. If the energy dissipation is satisfactory over the desired range of discharges and tailwater elevations, erosion tests may be performed on the downstream channel.

5–8. **Erosion Studies.**–Erosion studies, which give qualitative results, are made to compare the performance of various stilling basin designs. In making these studies, a sand, gravel, or cement-sand mortar capable of being readily moved by the flowing water is used to represent the channel downstream from the stilling basin. The measured depths and extent of the erosion will indicate the relative effectiveness of the basin, provided the same movable bed material is used for each test. Erosion depths may or may not scale to the prototype, depending on the choice of the model bed material; however, the model, when operated for a period of time sufficient to produce a stable bed, will indicate erosion tendencies and patterns in the prototype.

Model tests made using a movable bed of a lightly cemented cement-sand mixture may closely represent the prototype erosion if the eroding velocity of the prototype bed material is accurately estimated. To determine the effect of an incorrect estimate of the critical erosion velocity, other tests similar to the one just described should be made for velocities both double and half the first estimated value. If the erosion produced by the three tests is acceptable, no further tests need be made. Should one or more of the tests indicate excessive erosion, changes in the stilling basin or other parts of the structure may be necessary.

5-9. Measurement of Pressure, Velocity, and Turbulence.–Measurements of pressure are made on various parts of an energy dissipator (1) to determine whether pressures conducive to cavitation or vibration will occur on baffle piers, sills, or other parts of the structure exposed to high-velocity flow, and (2) to determine the forces on the structure caused by dynamic (fluctuating) pressures [18].

In stilling basins having high-velocity flow, dynamic pressures on walls, blocks, and sills should be measured. Pressure distribution should be measured to ensure against excessive subatmospheric and unsteady pressures where curvature of the floor or walls is used to guide or spread the flow entering the basin.

Dynamic pressures on the walls or the floor of stilling basins should be measured to help the designer determine unbalanced loads on the structure. Unbalanced loads caused by large, rapidly changing flow depths may occur for symmetrical or unsymmetrical operation. The investigator should anticipate unusual operating conditions in the prototype and make measurements that are of the most use to the designer.

Satisfactory measurement of pressures may require altering standard model construction methods. For example, where wood will normally suffice for construction of baffle blocks, chutes, or walls, metal construction may be required to permit precise piezometer installation.

Other observations that should be made in stilling basin studies include measurement of eddy velocities or local velocity concentrations, water surface profiles along training walls or other parts of the structure, and measurement of wave and surge heights in the downstream channel. Wave and surge height measurements are particularly important when the discharge is passed directly from the basin into an unlined channel because of possible destructive erosion of the banks. Readings from staff or point gages are often used to determine wave and surge heights; however, high-speed movies have shown that direct ۱eadings give less height than actually occurs, particularly when the wave period is short. Thus, when more accurate model prototype comparisons of the heights and periods of waves are desired, data should be recorded by movies, oscillograph and analog wave sensors, or by probability probe instruments.

D. OUTLET WORKS

5–10. **General.**–An outlet works may have single or multiple conduits, and usually includes valves or gates for flow regulation and an energy-dissipating structure. The various types of valves and gates are discussed in detail in Chapter 9, Control of Flow. The type of outlet valve or gate to be used is selected by considering (1) cost and (2) the desired prototype performance which in conjunction with the stilling basin structure provides satisfactory energy dissipation.

Valves and gates for use in outlet works may be classified as: (1) those that produce a solid or converging jet and (2) those that produce a diverging or hollow jet. Either category usually requires different types of energy dissipating structures than the other.

Slide gates of various types produce a solid jet thaᵗ is usually discharged onto a sloping chute which terminates in a conventional hydraulic jump basin, or they may discharge into a rectangular or circular conduit leading to an energy dissipator downstream. These gates are used on both high- and low-pressure outlets.

Hollow-jet valves may discharge onto sloping chutes, into tunnels, or into specially designed energy dissipators. Diverging or hollow-cone valves may be used in free discharge, in tunnels, and in specially designed tunnel energy dissipators.

5–11. **Model Design.**–Models of outlet works should ideally be geometrically and dynamically similar to the prototype. However, because the absolute roughness of the prototype conduit surface is nearly the same as that of the model, resulting in greatly differing relative roughness ratios, friction losses are disproportionately greater in the model than in the prototype. Thus, it is necessary to alter the geometric similarity to attain prototype dynamic conditions. For investigations that include entrance or intake studies and long conduits, the scaled prototype pressures and velocities can be produced in the model by making the conduit shorter than indicated for geometric similitude.

In many investigations, entrance studies are considered to be unnecessary, so only a short length of conduit is provided immediately upstream from the gate or valve. Figure 5-12 shows such a model in operation. The approach conduit, of sufficient length (10 or more pipe diameters) to produce a turbulent velocity distribution, may be connected directly to a water supply source. Piezometers are usually installed in the conduit one diameter upstream from the valve for measuring the pressure head on the valve. Water is then supplied to this point to represent the prototype flow rate, and the model valve is regulated to give the computed pressure at the piezometers as estimated by analytical studies and computations to represent the resistance of the prototype conduit. Thus, the model valve discharge will properly represent the form of the jet and the energy delivered to the stilling basin of the prototype.

Figure 5-13 shows a model of an outlet works using top-seal radial gates for flow control. Part of the upstream and all of the downstream lengths of conduit were constructed in the model.

Experience has indicated that model gates and valves should be at least 75-millimeters (nominal) diameter for satisfactory study of an outlet works energy dissipator. Smaller valves require special care in operation to ensure accurate results. Furthermore, 75-millimeter or larger valves are capable of discharging sufficient water to make observers appreciate the problems to be investigated.

Tests on outlet works stilling basins should be made as described in the general discussion on stilling basins. Because the jet from a gate or valve is generally concentrated, large fluctuating pressures are generated in the energy dissipator. Measurements should include the fluctuations in pressure in relation to the stability of the structure components; vibration and/or acceleration measurements on parts of the model, such as training walls, may be necessary. Wave suppressors near the downstream end of the basin may be used to reduce wave heights at the entrances to canals or tunnels [17]. Velocity concentrations at basin exits should be avoided to minimize erosion of the downstream channel. Also, flow concentrations and eddies tend to draw material into the channel and energy dissipator.

Figure 5-12.–Model of outlet works with hollow-jet valves. Top photo P801-D-79222, bottom photo P801-D-79221

Outlet works. P801-D-74605

Gate structure. P801-D-74600

Figure 5-13.–Model of top seal radial gate outlet works. P801-D-74600

99

CHAPTER 6

Canal Structures and River Channels

A. CANAL STRUCTURES

6–1. Introduction. –Canals may ordinarily be designed by analytical methods because they are usually uniform in cross section and slope. However, canal structures such as drops, overchutes, turnouts, wasteways, river crossings, or combinations of these may present design problems which are not readily solved by analytical methods. For such structures, the laboratory can assist in producing a design that gives satisfactory performance. Hydraulic models of canal structures, like those of spillways and outlet works, are scaled geometrically according to the Froude law because gravity is the predominant force. In models where friction forces are significant, such as in long, flat chutes, consideration should be given to adjustment of the model slope or roughness to compensate for the extra head loss.

Many canal structures dissipate energy in a manner similar to spillways and stilling basins; consequently, no attempt will be made here to repeat the criteria for study of rectangular chutes and stilling basins, pressure measurements, or other subjects already discussed. However, another useful energy dissipator, the impact-type basin, has been developed specifically for releasing water from canals and small reservoirs and it is described in the following section.

6–2. Impact-Type Basin.–An impact-type stilling basin (fig. 6-1) is an effective energy dissipating structure that requires no tailwater for successful performance. Dissipating the energy of an incoming jet from a pipe or channel is accomplished by impact of the jet on a vertical baffle and by eddies resulting from the impact. The basin can be constructed in various sizes to accommodate a wide range of flow rates [19]: as small as 250 liters per second from 300-mm-diameter pipes discharging into basins 1 to 1.25 meters

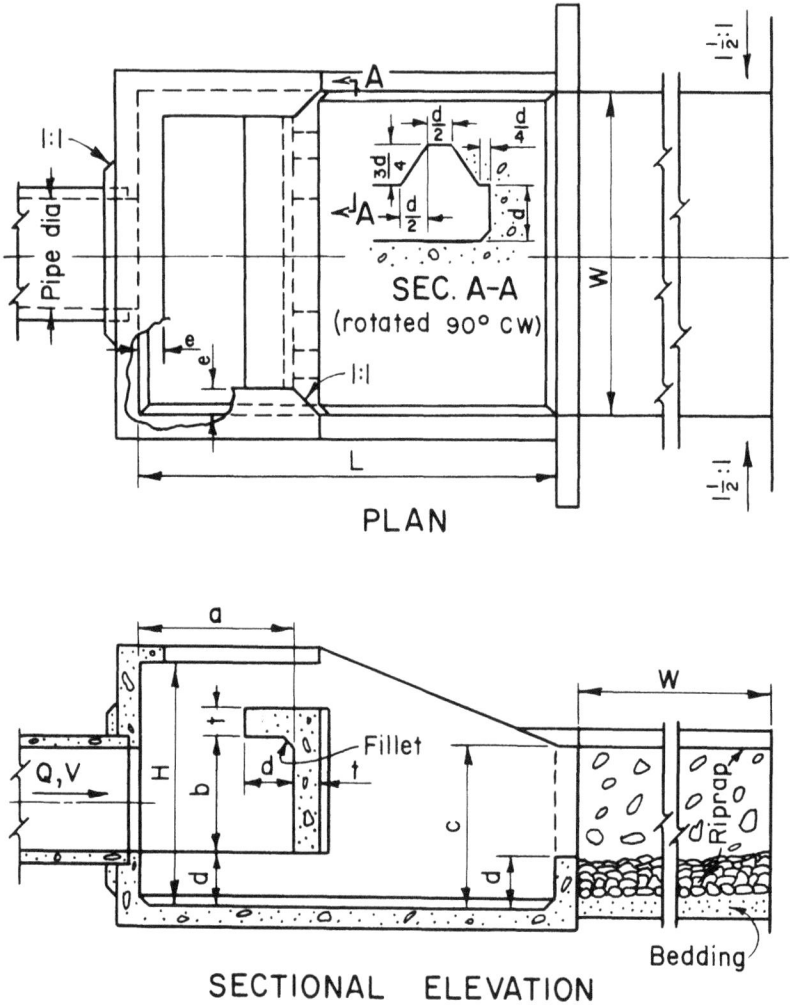

PLAN

SEC. A-A
(rotated 90° CW)

SECTIONAL ELEVATION

H = ³/₄ (W)   d = ¹/₆ (W)
L = ⁴/₃ (W)   e = ¹/₁₂ (W)
a = ¹/₂ (W)   t = ¹/₁₂ (W), suggested minimum
b = ³/₈ (W)   Riprap stone size diameter = ¹/₂₀ (W),
c = ¹/₂ (W)

Figure 6-1.–Impact stilling basin. 103-D-1719.

wide to in excess of 25 cubic meters per second from 1500-mm-diameter pipes into basins 9 meters wide. Velocities of flow from the pipe should be limited to a maximum of 15 meters per second.

6–3.   **Chutes and Overchutes.**–(a) *Chutes.*–Chutes (often called drops) are used to abruptly change a canal level or to deliver water to a wasteway. The chute conducts the water from the upper level through a stilling pool and transition to the lower level. A check structure in the canal is sometimes used to maintain normal depth in the canal upstream from the chute. Checks are designed in many forms; they may be weirs, gates, stoplogs, specially designed notches, or mechanical schemes for regulating the flow. A check and chute, trapezoidal in cross section, are shown in figure 6-2. In developing a chute structure, the check should be carefully calibrated to ensure that the required amount of water will be passed for all elevations of the upstream water surface. A curve of discharge versus water surface elevation will provide the information necessary to determine if the design will meet this requirement. Checks often produce unwanted wave formations in the chute. The chute and stilling pool should be proportioned to operate over the desired range with a minimum of surges and waves.

Figure 6-2.–Check and chute for trapezoidal canal. P801-D-79220

In drops that do not have sufficient tailwater depth for the stilling pool, a control or check may be used to increase the pool depth. The increased pool depth will help prevent high-velocity flow from entering the lower canal, creek, or natural drainage area. Stilling pools that cause waves and surges require special study because the waves may damage the canal downstream. Designing the pool to reduce waves and surges is usually more economical than providing bank protection such as riprap or lining. Baffle piers or sills may help reduce wave action, but the possibility of their effectiveness being reduced by the accumulation of weeds or debris should be considered.

The transition from the stilling pool to the canal should be structurally simple and not have a large angle of divergence. Rapid expansions promote unstable flow and magnify surface waves and surges. Eddies produced by improperly designed transitions tend to undermine the ends of the transition walls.

Because most canals are trapezoidal in cross section, a stilling basin of the same shape eliminates the need for complicated transition sections or wing walls at the end of the stilling basin. Also, the concrete can be placed with a minimum amount of reinforcing steel and formwork, representing a considerable saving in materials and construction costs over the conventional rectangular basin.

Trapezoidal shapes have the inherent characteristic of producing an uneven distribution of flow in a stilling pool. The flow concentrates and shoots through the center of the basin, leaving pronounced eddies on each side of the basin. Early model studies of a trapezoidal drop structure for a wasteway on the Boise Project in Idaho showed that a fairly uniform flow distribution could be obtained in the basin by placing longitudinal ridges along the bottom of the chute and stilling basin. However, these studies were confined to flow through a single control notch before the water dropped into the trapezoidal basin. Therefore, to develop satisfactory stilling pool operation when the flow enters the drop structure through two or more notches comprising the control, model studies of a trapezoidal drop structure of this design were undertaken for the Wyoming Canal of the Riverton Project. The control was designed to maintain the water surface at normal elevation in the canal upstream from the drop. The model studies resulted in a drop design that provided a relatively uniform distribution of flow in the stilling basin and permitted the water to enter the downstream canal with a minimum of scour and waves.

(b) *Overchutes.*–Overchutes are used to carry runoff from washes and ravines over a canal. In arid regions many structures of

104

this type may be necessary for one canal. Without them, sediment and debris would find their way into the canal, and canal breaks might be caused by the flow overtopping the canal banks. The cost of removing debris or repairing break damage could be substantial. A typical overchute resembles a bridge over a canal, with a slab of concrete paving downstream, appropriate training walls, and an energy dissipating device. Model studies are generally concerned with determining the capacity and the effectiveness of dissipation.

An overchute is shown in figure 6-3. The flow passes over the bridge and then over sloped paving containing rows of baffle piers. Backfill is placed over the lower end of the paved apron to the level of the downstream channel. The piers are arranged for dissipating the energy in the flow passing over the piers at a rate nearly equal to the kinetic energy developed during free fall. Thus, as the water tumbles down the slope, the kinetic energy of the flow remains approximately constant. Moderate degradation of the downstream channel will expose additional slope and piers, and then the same erosion pattern occurs at a lower elevation.

6–4. Turnouts and River Crossings.–Turnouts serve to divert flow from a main canal into a lateral system. Designing this type

Figure 6-3.–Wash overchute structure and baffled apron drop. P801-D-79219

of structure may be difficult, particularly if high-velocity flow or bends exit near the turnout. Trashracks installed on turnouts to prevent floating weeds or debris from entering the canal laterals can cause operating difficulties. Hydraulic models can be used to solve a variety of turnout problems and to help assure that the required flow of water in the lateral will be obtained. Measurement of pressures in the turnout may be required if high-velocity flow occurs in the canal or turnout. Investigations may determine whether the turnout is likely to intercept weeds or debris floating in the canal. Model comparisons can demonstrate configurations that will intercept the least trash.

At locations where canals intersect rivers or other natural drainage channels, the cross-drainage flow may be carried beneath the canal in a culvert or the canal flow may be conveyed under the channel through an inverted siphon. Model studies can help select the type to use for a particular situation. Unusual entrance, pressure, velocity, bend, or exit problems may make model studies desirable for either type of structure.

6–5. **Conduit Transitions.**–Inlet or outlet transitions are used to guide the flow between canals and siphons or tunnels in such a way as to minimize turbulence and wave action. Flow velocity distribution, wave generation, and head loss may have to be studied to develop a satisfactory structure. Descriptions and hydraulic model results of prior laboratory investigations may assist in planning new studies [20].

B. RIVER CHANNELS

6–6. **General.**–The term "river channel model" is used in a broad sense and applies to models in which the slope of the water surface throughout the model is relatively flat. These models usually involve studies of flow patterns or movements of sediment and are of two types: fixed-bed and movable-bed models.

Studies utilizing river models include:

- Effects of closing or opening certain channels in a multiple-channel river,
- Hydraulic losses at constrictions such as bridge piers or cofferdams,
- Probable erosion around cofferdams or other structures,
- Characteristics of flow in and around diversion schemes,
- Effects of currents and waves on river navigation,
- Tidal and estuary channels, and
- Ice effects on riverflows.

106

6–7. **Fixed-Bed Models.**–Canals or rivers having relatively long stretches of stable bed configuration are usually studied with fixed-bed models (fig. 6-4). Studies of the channel include backwater effects caused by obstructions or channel improvements, flood routing, determination of flow distribution in estuary channels, and ice flows. Special attention must be given to the model boundary resistance so that the Froude law of similitude may be applied. For large models, the velocity may be high enough to ensure turbulent flow and the minimum model roughness may have smooth enough boundaries to properly represent the prototype, while small models may not yield satisfactory results because boundary surface roughness prevents a scaled velocity.

Small models require a distortion in vertical scale and/or slope to (a) offset the disproportionately high resistance of the model boundaries and (b) obtain a sufficiently high value of Reynolds number to ensure turbulent flow. The distortion in slope required

Figure 6-4.–Fixed-bed model of river for study of booms used to control location of ice jams. P801-D-79244

107

to satisfy condition (a) can be computed with sufficient precision by the Manning equation:

$$V = \frac{1}{n} R^{2/3} S^{1/2}$$

Because Froude similitude shows the velocity ratio, $V_r = D_r^{1/2}$, the formula may be written

$$V_r = \sqrt{D_r} = \frac{R_r^{2/3} S_r^{1/2}}{n_r}$$

Substituting D_r / L_r for S results in the expression

$$\sqrt{D_r} = \frac{R_r^{2/3} \sqrt{D_r}}{n_r \sqrt{L_r}}$$

This expression may be rearranged as follows:

$$\frac{D_r}{L_r} = \frac{n_r^2 D_r}{R_r^{4/3}} \tag{1}$$

If the roughness coefficient, n, can be determined for model and prototype, then n_r is known and the distortion D_r / L_r can be computed for the hydraulic radius ratio of the mean section.

When the slope distortion required to satisfy condition (b) is greater than that required for condition (a), the model boundaries must be made rougher by artificial means to compensate for the exaggerated slope.

The required model roughness can be computed from equation (1) if the distortion and the prototype n are known. Model boundaries can be roughened by distributing bits of mortar, grains of sand, or gravel on the flow surface. Model roughness can be designed when general data on the equivalent roughness k of materials such as screen or expanded metal lath are available. If such information is not available, the materials must be formed and placed in the model and the head loss measured. This procedure may have to be repeated to obtain the proper roughness.

A model with roughness adjusted for a particular depth will yield dependable results for flow at or near this depth, but when a problem involves several depths, the model roughness should be adjusted to give an average resistance that is approximately correct over the desired range, or the roughness may be varied with depth for a closer approximation. One method of varying the roughness is to place sections of wire screen in the channel. The sections of screen, folded in a zigzag pattern, will stand by themselves and can be moved about to provide the appropriate resistance to the flow, thus simulating the correct roughness.

When the surface roughness of the prototype flow boundary is unknown, studies are made for a range of model resistance which is estimated to encompass the resistance caused by the prototype roughness.

River models have long been used in the study of ice formation and its control. In 1918, an investigator used a hydraulic model, in which he simulated float ice with paraffin, to study ice diversion. Other investigators later used hydraulic models to arrive at a better understanding of ice processes on rivers and reservoirs. Materials such as wax, paraffin, wood, and polyethylene were used to simulate ice in these studies.

There are two areas of similitude to be considered in modeling ice processes. One is concerned with the modeling of ice properties and the other embraces the modeling of individual ice flows in a river where the internal properties of ice are neglected and the similitude is based on hydrodynamic considerations. The two areas, however, are seldom completely independent of each other. One area usually dominates, and in the ice-cover formation on rivers, hydrodynamic considerations may dominate the formation process.

6–8. **Movable-Bed Models.**–Limited studies of open-channel problems involving scour, deposit, and transport of channel-bed material are made by investigating current direction and velocity in fixed-bed models; comprehensive studies require movable-bed models. Despite the limitations and roughness of modeling materials, movable-bed models prove to be a valuable aid in solving complex problems involving the shifting of streambed materials. Mathematical analysis for similitude in movable-bed models is not as rigorous as that applied to models of hydraulic structures having fixed boundaries. Because the four principal forces (pressure, gravity, viscous shear, and surface tension) cannot be simulated simultaneously in the movable-bed model, a successful study requires a balance of hydraulic forces to produce bed movement having the character of observed or expected prototype change.

The general design of the model requires selecting scales and materials that will result in bed movement generally similar to that in the prototype for the estimated discharge range. There are two prerequisites to such a design approach: (1) thorough knowledge of the characteristics of the prototype based on the collection and study of hydraulic and hydrographic data, and (2) experience in the field of movable-bed hydraulic models.

A nonconforming relationship is encountered when the dimensions of a watercourse, including the particle size of the bed material, are scaled down to model size. The usual model scales result in sediment particles that are so small they no longer act like bed material; instead, the material tends to become either suspended in the flow or compacted into an unyielding bed. Therefore, bed material of larger particle sizes is used to offset the scaling effect. Einstein and Chien [21] proposed a method to determine the similarity conditions for distorted river models with movable beds. Their method was based on the theoretical and empirical equations describing the hydraulics and the sediment transport processes in a river. Geometrical distortion of the model then becomes necessary to produce velocities sufficient to move the bed material. To obtain the necessary velocities, the vertical scale ratio is made larger than the horizontal ratio, exaggerating slopes in the model. The distortion (vertical scale ratio divided by horizontal scale ratio) ranges from 2 to 7, or even larger in special cases. Generally, distortion should be kept to as low a value as possible without appreciably reducing bed movement.

For investigations of sediment action relating to a structure such as the canal intake and sluiceway of a diversion dam, distortion which would include the structure is not advisable. In such studies, the model must be large enough to ensure movement of available model sediment to produce qualitative results [22].

With a movable-bed model, distortion of the scale ratio is usually necessary to obtain sufficient bed movement, but the distortion also introduces undesirable effects. The exaggeration may increase the slopes of the model banks beyond their angle of repose so that they cannot retain their shape. Distortion increases the longitudinal slope of the stream and tends to upset the flow regimen; artificial roughness may be required in the model to produce an acceptable flow regime. The vertical exaggeration also causes distortion of the lateral distribution of velocity and kinetic energy.

The difficulties of reaching an acceptable balance between boundary resistance and bed movement can be minimized by using model bed materials of lower specific gravity than present in prototype channels so that less scale distortion is required to produce proper movement. Among such materials are coal, pumice, sawdust, ground plastics, etc.

The apparatus that supplies material to a movable-bed or surface particulate model is usually located near the channel entrance. Rate of supply should be adjustable to permit duplication of sediment variation with prototype flow. One method of feeding bed materials developed in the Bureau's laboratory is described in Chapter 1, Laboratory Equipment. Another device is a submerged elevator that raises sediment to the channel bed. A third method is to use jet pumps or turbine pumps (if wear can be tolerated) to pick up sediment from a sump and recirculate a mixture of water and sediment to the model inlet (fig. 6-5). Surface particulate material, such as used in river ice studies, may be applied to models through a speed-controlled, fluted cylinder at the bottom of a triangular trough (fig. 6-6).

Figure 6-5.–Turbine-pump circulation system for 0.2-millimeter sediment. P163-D-24483

Figure 6-6.–Apparatus for supplying plastic beads to water surface to represent ice flow. P801-D-79246

6–9. **Verification of Movable-Bed Models.**–The verification of a movable-bed model is an intricate, "cut-and-try" process. Hydraulic forces and model-operating techniques are varied until the model will reproduce, with acceptable accuracy, changes in bed configuration known to have occurred in the prototype between certain dates. In this way, the accuracy of the functioning of the model is established. Certain scale ratios, such as time and discharge ratios, are determined experimentally. The verification procedure may consist of the following steps:

(1) Two prototype bed surveys of past dates are selected (the period between these two dates is known as the verification period). The model bed is molded to conform to the earlier survey.

(2) The hydraulic phenomena that occurred in the prototype during the verification period are simulated in the model to see if the events occurring in nature are reproduced in the model at the proper time. The time scale for the model is estimated.

(3) The model bed is surveyed at the end of the verification period. The model is considered to be satisfactorily verified only when this survey checks the prototype survey with acceptable agreement.

During the "cut-and-try" verification, manipulating the time scale, discharge scale, rate and manner of sediment feeding, or the slope scale of the water surface, and perhaps the gradation of the bed materials, may be necessary. Often, it is necessary to use scales for time and discharge varying with water depth so that each model flow depth will contribute to its proper share of movement. Therefore, the verification phase of a movable-bed model calls for an intimate knowledge of the prototype as well as experience with this type of model study.

After satisfactory verification of a river model has been made, the degree of similitude attained remains largely a matter of judgment. The similitude is based on this general reasoning: If the model accurately reproduces changes known to have occurred in the bed of the prototype, it can be relied on to predict changes of a similar nature which can be expected to occur in the future. Distortions in a movable-bed model place an important limitation on the type of tests that can be made and thus on the type of results that can be obtained from the study. Because the verification is achieved on the basis of an adjusted simulation of recorded prototype phenomena, the model cannot be expected to respond accurately to conditions which involve a drastic departure from those involved in its verification. In the final analysis, the validity of the results of a movable-bed river model study and the interpretation of results are largely dependent on general judgment and reasoning. The basis of such reasoning is the verification of the model, a knowledge of the prototype, and a familiarity with the general characteristics of such models (fig. 6-7).

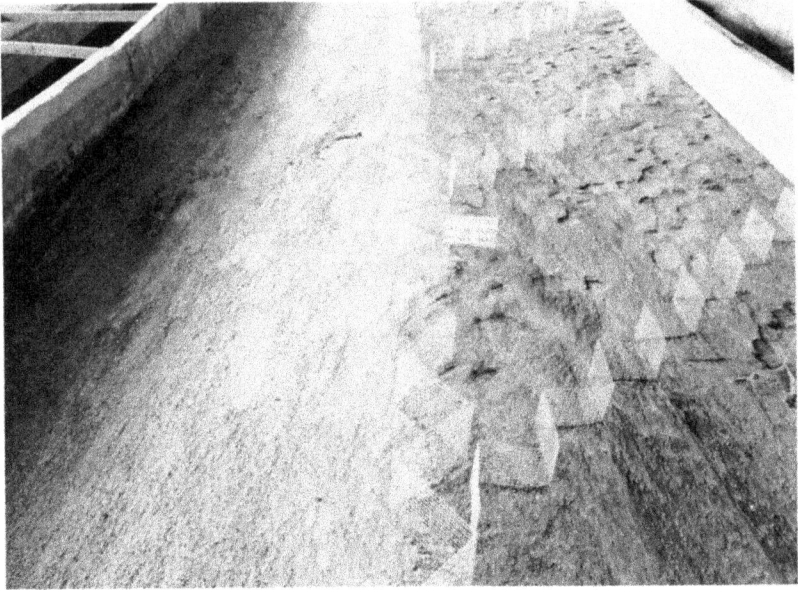

Figure 6-7.–River channelization study. Verifying artificial resistance at two scale ratios. Top photo P801-D-79217, bottom photo P801-D-79218

CHAPTER 7

Related Problems of Free-Surface Flow

7-1. **Fish Protection.**–Rivers are necessarily controlled to supply water for human needs. Because this control disturbs the natural habitat of fish, hydraulic investigations are justified in order to protect and, if possible, increase the numbers of fish for recreation and as a food source. Studies are made of such fish structures as ladders, louvers, and collection facilities. Ladders allow the fish to be self-transported from one level to a higher elevation around an obstruction. Louvers or racks are used to stop upstream migration and to guide the fish to collecting facilities. Collecting facilities trap the fish for scientific study or for transport to areas less likely to cause fatality.

One problem common to all types of fish structures involves the methods of introducing flow in such a way as to avoid high velocities and extreme turbulence. Hydraulic models can be very helpful for observing currents and turbulence and can confirm the design assumptions and principles of the structures.

Studies of fish ladders must take into consideration that salmon do not feed during their upstream migration, so they must live off the energy stored in their bodies. Thus, a relatively easy passage must be provided at obstructions so the salmon do not expend all their available energy before reaching the spawning grounds. Because they always swim into the flow, the current must be strong enough to induce them to move in the proper direction, that is, toward and up the fish ladder. To ease passage, the entrance to a fish structure such as a ladder should be placed adjacent to the obstruction. Turbulence from the flow spilling into the river from the ladder should attract the fish to the entrance and the turbulence created in the ladder by the water falling over weirs will direct the fish up the ladder. If the turbulence is excessive, however, the fish may jump out of the ladder. Also, if the velocity of flow in the ladder

115

is too great, the fish may expend too much energy. Should the exit of the ladder be placed too close to an outlet works or spillway intake structure, the fish leaving the ladder may be caught in the swiftly moving water approaching the intake and be swept back downstream.

Fish screens and traps are used to divert the fish from their usual path into a basin or other structure for the purpose of collection. As mentioned before, excessive turbulence and velocities should be avoided. This may be an easy matter when the racks are clean, but tests should be made to determine the effects of a rack partially clogged with debris. Because relatively few fish structures have been built and design information is meager, and the success of the structures depends on the habits of the fish, the advice of authorities on fish biology should be sought before finalizing a design.

Hydraulic model studies were essential to the design of the fish collecting facilities in the headworks of the Delta-Mendota Canal on the Central Valley Project in California. The facilities, developed after considerable study and experimentation, have prevented destruction of great numbers of game fish that would have otherwise been drawn into the large pumps of the project's Tracy Pumping Plant at the canal headworks. An interesting facet of the studies, rarely encountered in Bureau of Reclamation project development, was the influence of tidal fluctuations on the design and related laboratory investigations.

7-2. **Aquifers and Drainage.**–Agricultural production from irrigated lands diminishes over the years if adequate soil drainage does not occur naturally or is not provided artificially. Irrigation water is often applied in excess of the crop need and of the capacity of natural drainage, resulting in a rising water table. If the ground-water level reaches the root zone of the crop, two negative impacts occur: (1) oxygen in the soil, necessary for plant development, is displaced by the water, and (2) the water introduces dissolved salts detrimental to the health of the crop. Therefore, drainage systems are installed to collect and dispose of water in excess of the desired ground-water level.

During percolation of water and dissolved solids through the ground to the drain, additional salt may be dissolved from the soil. Disposal of the drainage flow back to the stream or river compounds salinity problems, particularly if salinity was already high at the point of irrigation water withdrawal. The problem of achieving a satisfactory salt balance may be difficult to solve, depending on the available water supply. Analytical and laboratory investigations provide information on the mechanism of ground-water movement, aiding in the design of efficient drainage systems. An important factor in the study of drainage systems is the resistance to flow caused by "convergence" as drainage flow approaches the envelope or gravel pack and enters the gravity or pump drain.

A laboratory study idealizes the cross section of an aquifer to linear scale in a glass-walled flume (fig. 7-1). The scale does not necessarily apply to particle sizes representing the aquifer nor to time factors used to interpret test results. Knowledge of the soil particle and aquifer sizes assists in selecting the differential permeability of the model material. Piezometers can be placed within the model aquifers or in the walls of the flume. Observation wells made of tubing are placed adjacent to the piezometer positions and are used to compare water depths and manometer-indicated pressures. A dyed, low-density saline solution is used to uniformly saturate the model aquifer [23].

Observations during a drainage study may include salt concentration in drain effluent, rate of flow into the flume and out of the drains, well levels, and piezometric head. Concentration of effluent samples may be analyzed by chemical or colorimetric methods. A flame photometer provides a rapid method of determining salinity values. Time-lapse movies at selected intervals provide graphic information on changes in drainage flow lines for study periods requiring 1, 2, or 3 weeks. Still photographs are valuable throughout the study.

A time scale for the investigation may be established indirectly. The prediction of prototype times from a scale model is not based on the usual Froude number relationship because of the lack of similarity with respect to particle sizes. Therefore, correlation of model and prototype times is made on the basis of comparative aquifer volumes:

A prototype aquifer 24 meters deep with a 40 percent total porosity would contain 24 x 0.4 = 9.6 cubic meters of water per meter of length. Deep percolation at 0.9 meter per year (a value obtained from a typical field project) would replace the original water content in a period of 9.6 ÷ 0.9 = 10.67 years. A 1:40 scale model, 4.79 meters long, 0.76 meter wide, and 0.60 meter deep, has a volume of 2.18 cubic meters. With a total porosity of 40 percent, the water content would be 2.18 x 0.4 = 0.872 cubic meter. An inflow of 12 liters per hour for an 8-hour period would supply 0.012 x 8 = 0.096 cubic meter. On this basis, the time required to replace the original water content in the model would be 0.872 ÷ 0.096 = 9.08 periods of 8 hours each. Thus, 6.81 hours (9.08 x 8 ÷ 10.67) of model time represent a year of prototype time for the recharge rate of 12 liters per hour to the model. This comparison is valid, however, only for an area irrigated over the entire surface at regular intervals or at a uniform steady rate.

117

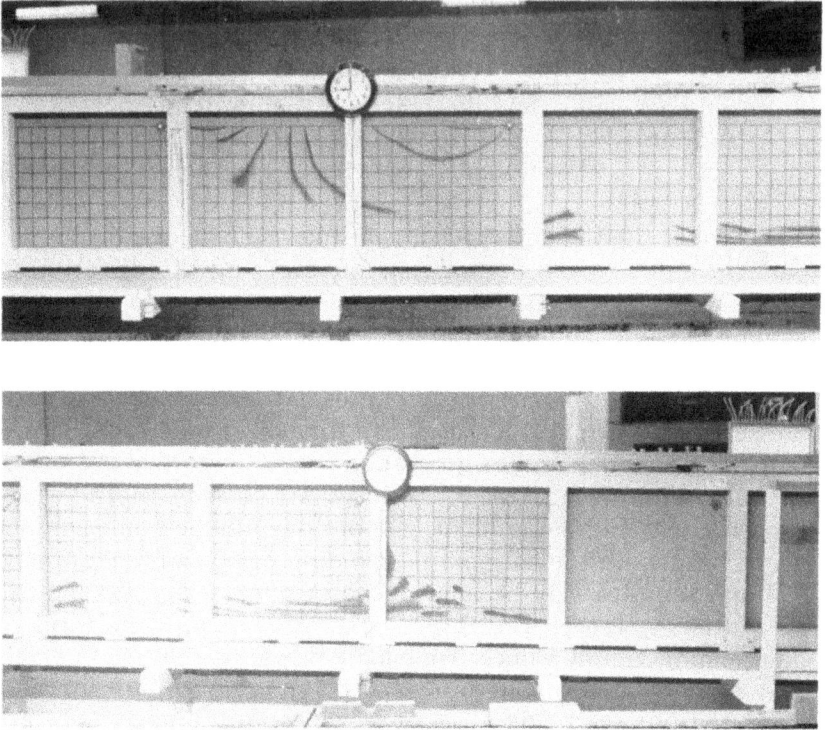

Figure 7-1.–Aquifer model showing flowlines toward drains traced by dye.
PX-D-70813 and P801-D-74584

A computed time required to replace water in the idealized
section of the agricultural area may be scaled to the time required
to replace the original water content in the laboratory aquifer. Part
of a day in the model may represent a year in the idealized
agricultural area.

In recent years, there has been increased use of sprinkler systems
to irrigate Bureau of Reclamation project lands. In contrast to the
methods of furrow or flood irrigation, which can be applied only
on flat lands, sprinkler systems can irrigate lands having relatively
steep slopes. Application of irrigation water to sloping lands has
compounded the need for drainage and has required expansion of
the basic knowledge in drainage technology.

Recognizing these facts, researchers in the Hydraulic Laboratory
undertook a comprehensive investigation to extend the knowledge
of the effect of land slope on depth and spacing of drains on irrigated
farmlands. The emphasis of the investigation was directed toward
determining what effect various slopes and irrigation rates would

have upon the location of the water table, drain discharges, and the path of water through the soil. Differing from the usual Bureau laboratory studies concerned with solution of specific design problems, this investigation had as its objective a broadened understanding of basic concepts and increased insight into the general principles of drainage. The results of these studies have led not only to the attainment of this objective, but they have also aided engineers in solving specific drainage problems and will continue to be of importance in designing new irrigation projects.

7-3. **Stratified Flow.**–Studies in density-stratified flow are made for the purpose of controlling water quality. The investigations include laboratory and field studies of idealized and natural rivers, reservoirs, and ground-water sources of water. Water temperature for maintaining or enhancing fish and wildlife, for irrigation, and for industrial uses may be controlled by using selective withdrawal from density stratification.

Dissolved oxygen necessary for biological degradation of sewage and industrial wastes and for preservation and growth of fish is related to density stratification. Density variations caused by dissolved salts and sediments affect tidal estuaries, coastal aquifers, and the life expectancy of reservoirs.

Mechanics of stratified flow are investigated in:

- Interfacial wave propagation, density currents, and siltation in docks and tidal basins,
- Turbulent mixing of dissimilar flows,
- Pollutant distribution in rivers,
- Motion of saline fronts in still water,
- Flow in saturated porous media,
- Saline water exchange from locks to freshwater channels,
- Cooling water circulation in canals, rivers, and reservoirs,
- Hydraulic jumps in multilayered systems,
- Atmospheric flow, and
- Selective withdrawal or exclusion in density-stratified rivers, lakes, and reservoirs.

Careful analysis is required in planning and performing meaningful investigations. Mathematical, electric analog, and physical models and prototype measurements are used in general investigations. Similitude relationships for stratified flows are being studied comprehensively. While theoretical solutions are being clarified, experimental results from previous studies guide the construction and operation of new studies. Model design includes the densimetric Froude number:

$$F' = \frac{V}{\sqrt{g'd}}$$

where

F' = densimetric Froude number,
V = average velocity in withdrawal layer,
g' = $g(\Delta \rho/\rho)$,
d = 1/2 the thickness of withdrawal layer, and
ρ = density of fluid

Both theoretical and experimental coefficients related to withdrawal layer thicknesses have been proposed for application to design of selective outlets [24].

Temperature or salinity differences may be used to induce density stratification in laboratory investigations. Thermocouple or thermistor temperature-measuring equipment is suitable for continuously monitoring changes in thermally stratified flows, while conductivity probes are used in salinity stratified flows. Sensors can be mounted on rods inserted from the top of the model or extended through the walls. Records of temperature change may be kept in analog or digital form, depending on the method of data analysis to be used. Time-lapse and still photography allow later interpretation and comparison of various schemes. Dye tracers within and across layers of flow improve visual interpretations of flow patterns. Measuring discharge rates and flow velocities, obtaining temperature or salinity samples from the flow, and determining water surface levels increase the complexity of the model. Both commercial and laboratory-developed equipment may be required to obtain all the necessary measurements. Automation of temperature or salinity controls, data acquisition, and analysis can be of advantage in long-period studies [25]. Figure 7-2 shows a test facility for selective withdrawal studies.

Field studies may include measurement of drain effluent turbidity, dissolved oxygen, pH, conductivity related to salinity, air and water temperatures, solar radiation, and water current velocity. Aeration of flows in reservoirs and rivers may be combined with studies of stratification [26].

7-4. **Flow Measurement.**–Measurement of water for irrigation, municipal and industrial uses, and for hydroelectric power development requires continuing study of measuring devices and techniques. Accuracy and dependability are the major criteria

Figure 7-2.–Test facility for selective withdrawal from a stratified liquid.
P801-D-79216

for judging the performance of a measuring method. A study should provide a complete description of the method or device and the procedures required for its use. Sufficient information should be acquired from the study to define advantages, disadvantages, and limitations of the device or method [5].

Laboratory investigations may involve standard or modified weirs, Parshall and trapezoidal flumes, orifices, commercial meters, and gates of various types. A structure that partially restricts flow in a channel can be used as a measuring device–if it is properly calibrated. Calibration of such a structure can be determined by thorough laboratory study.

Standard devices do not always satisfy the measuring conditions. Special methods or devices are then required, and their development by laboratory or field investigation is necessary. Measurement deficiencies encountered in use may be overcome by calibration of the device in place or by study in the laboratory.

Investigations require measuring head, discharge, velocity distribution, and flow losses in a manner similar to model studies. Evaluation of tracer, magnetic, ultrasonic, radioactive, and resonant methods of measuring may require use or development of special equipment. A reliable volumetric or gravimetric facility is necessary for comparing or calibrating the measuring method.

Weir box structures have been useful in measuring small flows of water to irrigated lands on Bureau of Reclamation projects. For example, irrigation turnout structures having a 0.91-m-wide suppressed rectangular weir for measuring discharges up to about

121

0.14 cubic meter per second have been in use for a number of years on the Columbia Basin Project in Washington. These turnouts have proved to be economical to build, easy to operate, accurate and reliable as flow measuring devices, and generally less troublesome than some other measuring structures. Because of the usefulness of the structure in gravity turnouts, additional studies were made of a 1.22-m-wide weir structure [27].

CHAPTER 8

Closed-Conduit Flow

8-1. **General.**–Laboratory investigations of closed conduits include a wide variety of hydraulic facilities, systems, equipment, and devices–ranging from large outlets, penstocks, or aqueducts to small fluid passages of machines or their controls. Although various closed hydraulic systems may serve different purposes, similar components of the systems involve investigations of a similar nature. In the design of a power penstock, minimizing hydraulic losses is desired; on the other hand, an energy dissipator requires that losses be at a maximum. Each study may concern one or more of the following hydraulic characteristics or conditions:

- Conduit size (based on efficiency or capacity),
- Entrance shape (based on pressures, losses, and capacity),
- Losses and flow action in pipe sections, bends, branches, fittings, and special shapes,
- Energy gradients and their relation to flow passage alinement,
- Effects of additions or alterations to existing systems (pressures, losses, and flow conditions),
- Design of flow-controlling device (with respect to capacity, efficiency, pressure distribution, losses, boundary shape, flow action, hydraulic practicability, and torque or other forces on operating mechanism), and
- Operating characteristics of hydraulic equipment and component parts (operating procedure, determination of capacity data for rating curves and tables, and flow action, including waterhammer and cavitation).

Many of these subjects will be covered in more detail in following discussions that deal with particular hydraulic systems, equipment, devices, and their component parts.

Model investigations of closed conduits are usually conducted using water as the working fluid. However, oil, air, or any other suitable fluid may be used. Of the fluids other than water, air is the most widely used in the laboratory. Testing with air offers many advantages in certain types of investigations.

8-2. Testing With Air as a Fluid.–Use of low-velocity air for studies of hydraulic equipment such as valves, gates, and other closed-conduit devices offers quick, inexpensive solutions to many hydraulic problems. A model can be constructed and tested more rapidly and with a degree of accuracy equal to that of a model that uses water. If air velocities are kept below 100 meters per second (about one-fourth sonic speed), the effects of compressibility can be ignored, and computations can be made using hydraulic formulas. If the results are expressed in dimensionless terms, they can be applied directly to the prototype. Figure 1-13 shows a laboratory centrifugal blower supplying air to a model. Low-cost, lightweight material can be used for the test structure, so modifications can be made readily. Modeling clay, molding plaster, or wood serve well for rapid changing of boundaries in cut-and-try experiments. In making tests with low-velocity air, the shaping of model surfaces or boundaries for constraining flow must be done carefully. Otherwise, results are likely to be in error and of little value.

Air in place of water as a test fluid is primarily useful for studies of flow in closed conduits and around submerged shapes. A minimum of equipment is required, and there is no problem of storing and disposing of the fluid unless smoke having objectionable properties is used to enable observation of flow action. Moreover, there are no difficulties encountered in bleeding piezometer tubes. Greater measuring precision might be required with air because the liquid columns used to indicate pressure or velocity are usually not more than 100 millimeters in height. Instruments (sloping tube, differential, and null-type manometers; pressure transducers; etc.) are available for precision measurements in air models.

Employing sectional models may be advantageous in air testing. These are usually sectors of cylindrical valves or similar devices, from which the air passes directly into the atmosphere. Sector models of less than 180° are most common.

If the whole structure is to be modeled, many of the component parts may be turned from wood, molded from plastic, or machined from plastic castings. If the sector-type model is used, molding plaster may be shaped through a process of building up and scraping off the surplus plaster by means of sliding or revolving templates.

8-3. **Cavitation.**–Cavitation occurs in a hydraulic structure when the pressure at a point in the water is reduced to the vapor pressure (about 3.5 kPa absolute for water). Cavitation takes place in this low-pressure region and consists of the formation, transportation, and collapse of vapor cavities. The cavities form in the regions of vapor pressure and are carried downstream where they collapse or implode as they reach a zone of higher pressure. The forces that accompany implosions which occur on solid boundary surfaces cause disintegration of the boundary material. This destruction is termed "cavitation erosion." (See fig. 8-1.) Severe cavitation accompanied by vibration is also likely to cause unstable operating conditions.

Laboratory studies help in developing cavitation-free designs by locating regions of low pressure. The model can then be modified to eliminate these low-pressure regions. Studies are also made of existing structures to eliminate or minimize vibration and damage from cavitation.

Cavitation in structures may be induced in many ways. The following causes are encountered most frequently:

- Curving a boundary surface too rapidly away from the normal path of a high-velocity fluid stream,
- Permitting irregularities or discontinuities to exist in boundary surfaces subject to high-velocity flow,
- Using extreme variations in elevation, resulting in a siphonic action in a conduit system,
- Moving unstreamlined objects through liquids at high speeds or passing high-velocity flow over unstreamlined objects, and
- Expanding the flow area too rapidly in the direction of motion.

Cavitation may produce severe destructive action in a structure, requiring corrective treatment or restrictions on the range of operation. One or more of the following measures might be applied:

- Impose strict operating limitations,
- Supply air to the low-pressure region,
- Alter all or a portion of the flow passage,
- Streamline parts directly in contact with high-velocity flow, or
- Maintain the damaged portion by periodically placing cavitation-resistant material in the eroded areas.

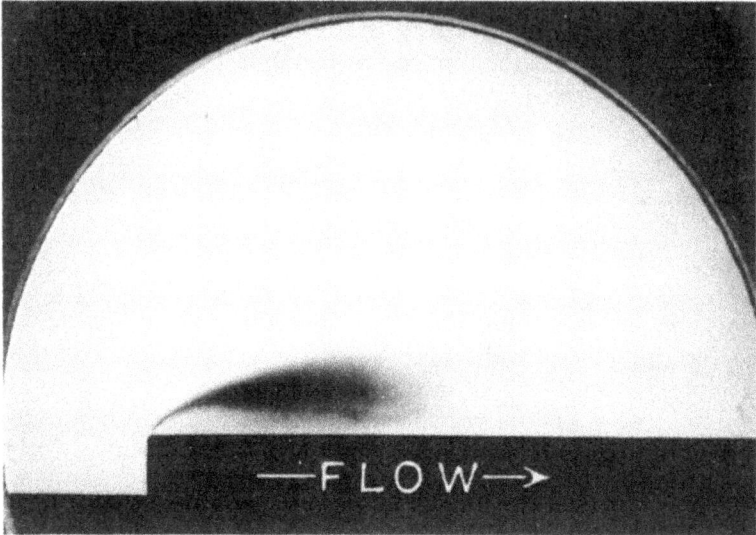

A.   Shadow of vapor cloud (cavitation) downstream from a 6-millimeter flow-surface offset.

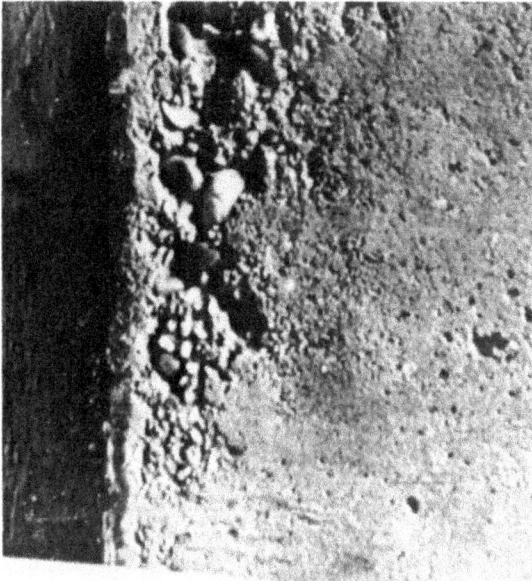

B.   Erosion of concrete resulting from cavitation caused by a 6-millimeter offset between steel gate frame and concrete guide wall.

Figure 8-1.–Examples of cavitation. P456-D-28059

(a) *Preventing Cavitation in Structures.*–Because much uncertainty exists regarding basic relationships among the hydraulic properties involved, designing cavitation-free structures through calculations alone is seldom possible. Where the validity of the calculation assumptions is in question, pressure profiles should be studied in scale models or special apparatus constructed for that purpose. The reliability of the information obtained from the laboratory investigation used in predicting prototype performance will depend on the suitability of the model, the accuracy and appropriateness of the data taken from it, and the rationality of the analysis and interpretation of the data.

High-velocity jets discharging through slide gates into lined tunnels and chutes at outlet works installations have caused serious cavitation erosion on several Bureau projects. To determine methods of preventing cavitation and to provide recommendations for altering existing structures and for designing new structures, the Hydraulic Laboratory conducted model studies of chute offsets, air slots, and deflectors.

In carrying out these studies, the Laboratory's investigators used a single test facility to model the structures at two existing dams, Palisades in Idaho and Navajo in New Mexico, and those at three proposed dams. To prevent cavitation erosion, air was introduced along the underside and sides of the jet from the gate before the jet came in contact with downstream concrete surfaces. Wall air-vent slots combined with a floor deflector were developed for use immediately downstream from the gate frames in the outlet facilities at Palisades and Navajo Dams. Wall and floor air-vent offsets away from the flow at the ends of the frames were developed for the new structures. These investigations, supplemented by general research, formed the basis for guidelines in the design of subsequent air-entraining devices to protect flow surfaces from cavitation erosion.

An example of a technique to prevent cavitation in a closed conduit is the aeration device in the spillway tunnel of Yellowtail Dam in Montana. The device was installed during repair and modification of the tunnel, concurrently with hydraulic model studies. The device–a nearly peripheral slot in the tunnel lining which admits air to the jet as it passes over the slot–entrains air in the water flowing through the tunnel, thus preventing cavitation damage to the concrete lining of the tunnel. In Bureau of Reclamation practice, the experiences derived from both the hydraulic model studies and prototype tests for the Yellowtail Dam installation serve as guides in the design of aeration devices for modification of existing structures.

(b) *Cavitation Investigations.*–Cavitation studies involve extensive measurements of pressure intensities and distribution on flow boundary surfaces. There are two techniques for testing geometrically similar models: (1) operation at atmospheric pressure; and (2) operation at scaled atmosphere or reduced pressure. When the problem involves prediction or elimination of cavitation, the first technique is suitable. Detailed measurements of pressure distribution in the model may reveal subatmospheric pressures. These pressures scaled to prototype values may approach the vapor pressure of water and be regarded as sufficient evidence of cavitation potential in the prototype.

For example, tests of a model valve indicate subatmospheric pressures of 2 meters of water at a model head of 20 meters. If the prototype head is 100 meters, subatmospheric pressures of 10 meters may be predicted, and cavitation in the prototype is possible. The model shape should then be altered to eliminate the subatmospheric pressures. Vacuums up to one-half atmosphere in the prototype are tolerated in special cases involving large-scale models. Subatmospheric pressures greater than one-half atmosphere are considered to be in the realm of potential cavitation. Irregularities in the boundary surface of the prototype may create local reductions in pressure and pressure fluctuations not detectable in the model studies; thus, subatmospheric pressures should be acceptable only in prototypes that can be constructed to very close tolerances.

Cavitation may not actually be produced in the model using water as the test fluid. In this case, the location of piezometers is of major importance. Piezometers for measuring pressure distribution must be located in critical areas selected on the basis of previous experience or analytical studies. In general, the piezometer openings should be located in expanding sections, curved surfaces, and downstream from discontinuities.

The fact that cavitation will occur can be ascertained from a model using water or air as the fluid, but the extent of cavitation and cavitation erosion cannot be accurately predicted from model pressure measurements. Procedures for analyzing pressure data depend on the scale ratios of heads used in testing. Prototype pressures for a given head can be calculated by multiplying the recorded model pressures by the scale ratio (Froude law). For general analysis, pressure factors and cavitation numbers (ratio of pressure at a point to the total head on the system) may be established from the model tests. The cavitation number may be defined in several ways; one is:

$$K = \frac{P_x - P_v}{\dfrac{V^2}{2g}}$$

where

P_x is the pressure at a particular point in the flow system,
P_v is the vapor pressure for the flowing fluid, and
V is the velocity at some reference point.

The relationship for K obtained from a model can be used to predict the cavitation tendency of the prototype structure. Pressure distribution on a surface of the full-size structure can be predicted with a high degree of accuracy if the scaled values indicate pressures above the vapor pressure of the fluid. On the other hand, if pressures below the vapor pressure are indicated, the results cannot be scaled accurately. Vacuum-tank testing may be useful when such conditions are present.

Vacuum-tank study of cavitation is complicated, but the tank simplifies the determination as to whether cavitation will occur and what the pressure distribution on a surface will be in the presence of cavitation. Using a model scale of 1:14 and an atmospheric pressure of 8.5 meters of water, the pressure in the tank for model atmospheric pressure would be about 0.6 meter of water absolute. Leakage of air into the system is an inherent problem in this type of testing, and pressure measurements may be more difficult because of vaporization and the special equipment required.

Vibration and energy loss resulting from cavitation can best be studied in reduced pressure models. These tests are performed in a water tunnel or in a closed system wherein subatmospheric pressures of various magnitudes can be maintained at the test section. Such a test facility makes possible the partial duplication of cavitation conditions by arranging equal values of the cavitation numbers of the model and prototype. For the same values of K in the model and in the prototype, the patterns of cavitation will be the same. Differences between the structural behavior of the model and the prototype will be large. Thus, careful attention must be given in transferring pressure fluctuation, vibration, and erosion damage from model to prototype [28].

(c) *Resistance of Materials to Cavitation.*–Special, easily eroded materials may be used on hydraulic models to show areas that will be damaged by cavitation and the relative extent of the affected surface.

Damage to materials by cavitation can be evaluated in the laboratory only in a relative way–among several samples of the same material or comparing one material with another. Erosion by cavitation has been used extensively to determine the relative abrasion resistances of engineering materials. For metals, cavitation by vibration of the specimen (magnetostriction oscillator) is an effective means of producing damage and showing relative resistance of the material. Engineering materials have been compared for resistance to cavitation erosion in venturi-shaped passages, paint coatings in cavitation cones constructed within transparent pipe, and concrete specimens in rectangular venturis. Submerged rotating disks with small obstructions in or on them have been used successfully to induce cavitation erosion.

8–4. **Flow in Pipes.**–Designing conduits for conveying water involves determining hydraulic losses, measuring the quantities of flow, and determining pressure intensities at particular points. Hydraulic losses are usually determined by piezometric head measurements in a prototype conduit system or by application of accepted formulas and known coefficients. Thus, loss determinations do not, as a rule, involve extensive laboratory studies, but may involve field studies. The Bureau of Reclamation has published a monograph on friction for concrete and steel pipes constructed by various methods [29].

Laboratory tests may be required for special studies, such as for unusual fittings, changes in section, or energy-dissipating devices. If a complicated system is involved, the laboratory may construct a schematic model of the system to simplify the investigation.

8–5. **Pipeline Surging.**–In pipe distribution systems that are provided with overflow standpipes, periodic variation of flow, or surging, may prevent delivery of water. The standpipes provide safety to the line from high static and waterhammer pressures, but they may become a source for initiating surging. The predominant initiating source of surging is air entrained by water as it falls over the baffles or crowns of the standpipe. This air collects in large bubbles in the pipe downstream from the structure, buoyancy of the bubble overcomes the force of the flow, and blowback occurs at the standpipe. A coincidence of the blowback and the natural period of the pipeline may amplify the surge, preventing delivery and overtopping the standpipe [30].

Investigations should include air entrainment tendencies, presence of initiating oscillations, and amplification possibilities. Time-synchronized pressure-recording, photography, continuous flow measurements, and visual observations are useful in acquiring data. Models of the pipeline are useful tools in separating parameters, but prototype operation and observation may be necessary for final analysis. Mathematical models are available for analysis of hydraulic transients in closed-conduit and open-channel flows [31].

8–6. Flowmeters.–Flow measurement devices or methods for closed conduits may be classified as:

(1) Those depending on the dynamic force of the fluid for their operation;

(2) Those measuring by volume or by mass; and

(3) Those using sound, magnetism, light, heat, or dilution of chemicals.

Devices using the dynamic force of the fluid for their operation include venturi meters, orifice meters, pitot tubes, flow nozzles, propeller meters, and their variations. Flow indicating and totalizing meters operating on these principles are used in irrigation practice. Volumetric meters are used principally for measurement of small quantities and are found in scientific instruments, gasoline pumps, domestic water and gas meters, and other similar devices. Meters indicating the discharge rate through application of an external physical quantity include ultrasonic, magnetic, thermal, and radioactive methods.

Laboratory studies on meters include calibration under various operating conditions and determination of losses, coefficients, limitations, and applicability. The meter should be located to discharge directly into a volumetric or gravimetric calibration tank or through a volumetrically calibrated measuring device. Test procedures and techniques are similar to those used in studies of other types of controls (chapter 9) and vary with the nature of the meter under study.

Flowmeters have numerous engineering applications in automation and control, but the laboratory is concerned mainly with flow measurements, which are used extensively in established formulas or model parameters in making computations from test data. Occasionally, the characteristics of a special orifice or nozzle design, or of a standard design placed in an unusual setting, are studied in the laboratory. Tests in such cases normally involve extensive discharge and pressure-differential measurements. Discharge quantities are commonly obtained from some standard or volumetrically calibrated laboratory device. Differential pressures across the device being tested are obtained with piezometer connections and a suitable recording gage. The differential pressure taps are important and should be placed to give consistent results. Locations of the pressure taps should be recorded with test data; otherwise the information loses its usefulness.

If orifice or nozzle losses are important, as in the case of the opening between a pipeline and a surge tank, sufficient piezometers should be provided to assure a complete and satisfactory analysis of the pressure. Pressure differential and head loss coefficients, expressed in terms of the velocity head, are useful parameters. Models of orifices and nozzles should be as large as practicable to assure flow conditions similar to those of the prototype. The approach and downstream velocity distributions should be reproduced carefully. In considering pressures and discharge capacity, the model is usually tested at scale heads. However, operation at prototype heads is frequently desirable, especially if subatmospheric pressures occur in the nozzle passage.

8–7. Siphons.–(a) *General.*–A siphon may be defined as a closed conduit, a part of which rises above the hydraulic gradeline. The negative pressure that prevails in a siphon gives rise to hydraulic problems that are complicated because two fluids–air and water–are involved. Siphons in automatic spillways and canal wasteways usually include an entrance, throat, barrel or siphon leg, and an outlet tube [32]. Siphon elbows, used in lieu of control valves at the ends of pump discharge lines, are special devices and are discussed in the following section. High efficiency, adequate capacity, stability of operation, and ease and time of priming are requisites of a good siphon.

The characteristic action of a siphon is to discharge the maximum flow whenever it operates. Consequently, the protective works at the discharge end of the siphon must be capable at all times of dissipating the energy of the maximum flow. A regulator installed in the siphon thus serves the purpose of regulating the siphon outflow so that the larger discharges occur only when necessary. The regulator also prevents or lessens the probability of damage at the outlet of the siphon.

The Froude law of similitude is applicable to siphons because the open-channel conditions at the entrance and exit have a large influence on the operation. Also, the conditions during the priming cycle are clearly governed by inertial forces which require observance of the Froude similitude relationships. Boundary resistance during normal full-flow operation does not conform to the Froude law, but the effects are minimized by making the model surfaces as smooth as possible. When the model results are extrapolated to the prototype, the disproportionately high resistance is adjusted through consideration of the Reynolds number and relative roughness. Occurrence of two-phase flow (air and water) during the priming cycle emphasizes the effects of surface tension, which can be minimized only by avoiding small models. Where the time of priming is critical, the influence of surface tension should be evaluated by using at least two models with different scale ratios.

(b) *Inverted Siphons.*–A conduit that dips to pass under a river or follows the ground surface across a valley is commonly called an inverted siphon. This type of siphon is filled by gravity flow and does not require priming because the flow of water replaces air as the system is placed in operation. Inverted siphons may be parts of closed systems or parts of what are otherwise open canal systems. The flow capacity of an inverted siphon running full, as affected by friction, exit, and entrance losses, may be computed from hydraulic relationships for flow in a closed conduit.

Laboratory studies involving models may be required for inverted siphons that are parts of canal systems; at partially full entrance and exit sections, air entrainment can be very troublesome. To avoid entraining excessive air, the entrance should be designed to give a minimum of turbulence, and flow should not plunge into the barrel. If plunging cannot be avoided, the siphon velocity should be sufficiently high to carry the air through the barrel without the formation of large bubbles, or it should be sufficiently low to allow the air to separate and discharge from the siphon at the upstream end [33].

High points in the alinement where air might collect should be avoided. Air collection tends to reduce the siphon capacity, and quantities of air released intermittently from high points results in objectionable turbulence. If the entrance runs full and other parts of the siphon immediately downstream are operating with a free water surface, sufficient venting should be provided to prevent reduced pressure at the water surface. A reduced pressure might result in the siphon entrance channel operating alternately as a closed conduit and as an open channel.

Model tests on siphons (Froude similitude) might include observation of the flow within the entrance (because the flow conditions affect the entrainment of air), the capacity of the system, and hydraulic losses of a local nature in various portions of the

133

structure. If the lowest portion of the siphon barrel is considerably lower than the entrance and exit, facilities capable of operating under high heads for draining the siphons are usually necessary. Such drain facilities are known as blowoff structures or vertical stilling wells.

Vertical stilling wells, a typical design of which is illustrated on figure 8-2, are ideally suited for dissipation of high-energy pipe flow. The pipe discharge enters the stilling well along a vertical axis through a control valve fastened to the floor of the well at the terminus of the pipe. The high-velocity jet which leaves the valve seat in a radial-horizontal pattern converges in the four corners of the well. Convergence of the radial flow results in very intense vertical flow in the corners. Corner fillets direct this vertical flow from the lower corners into the center of the well, creating a roller action which adds to the turbulence and energy dissipation. The flow rises vertically in the well where it is stilled and then discharged with a smooth water surface into a horizontal canal or chamber. The relative protection against abrasion and cavitation erosion provided by this type of energy dissipator and its potential to receive and still discharges from manifolded outlets more economically than a conventional stilling basin have been widely recognized.

One of the early (1950) Bureau of Reclamation laboratory studies of vertical stilling wells was for the Soap Lake Siphon, a segment of the West Canal on the Columbia Basin Project in the State of Washington. The study was valuable in setting the trend for later designs of vertical stilling wells. Although early studies were useful in developing specific stilling well designs, the economics of today do not justify such individual studies of small structures. For this reason, the Hydraulic Laboratory later (1962-1970) carried out an extensive program of research to develop design criteria covering a wide range of heads and discharges in stilling wells.

8-8. **Siphon Elbows in Pump Discharge Lines.**–Siphon elbows at the ends of pump discharge lines are used instead of valves or gates for the purpose of preventing backflow from the receiving reservoir during pump outage. During normal operation, the siphon elbow flows full, and a negative pressure prevails in the siphon bend or throat section. When the pumps are stopped, an air relief valve or siphon breaker admits air to the throat, raises the pressure, and disrupts backflow through the siphon. Design of these siphons has not been standardized, and further model and field studies are required [34]. The discussion of similitude relationships in the preceding section would apply in such studies, but there are

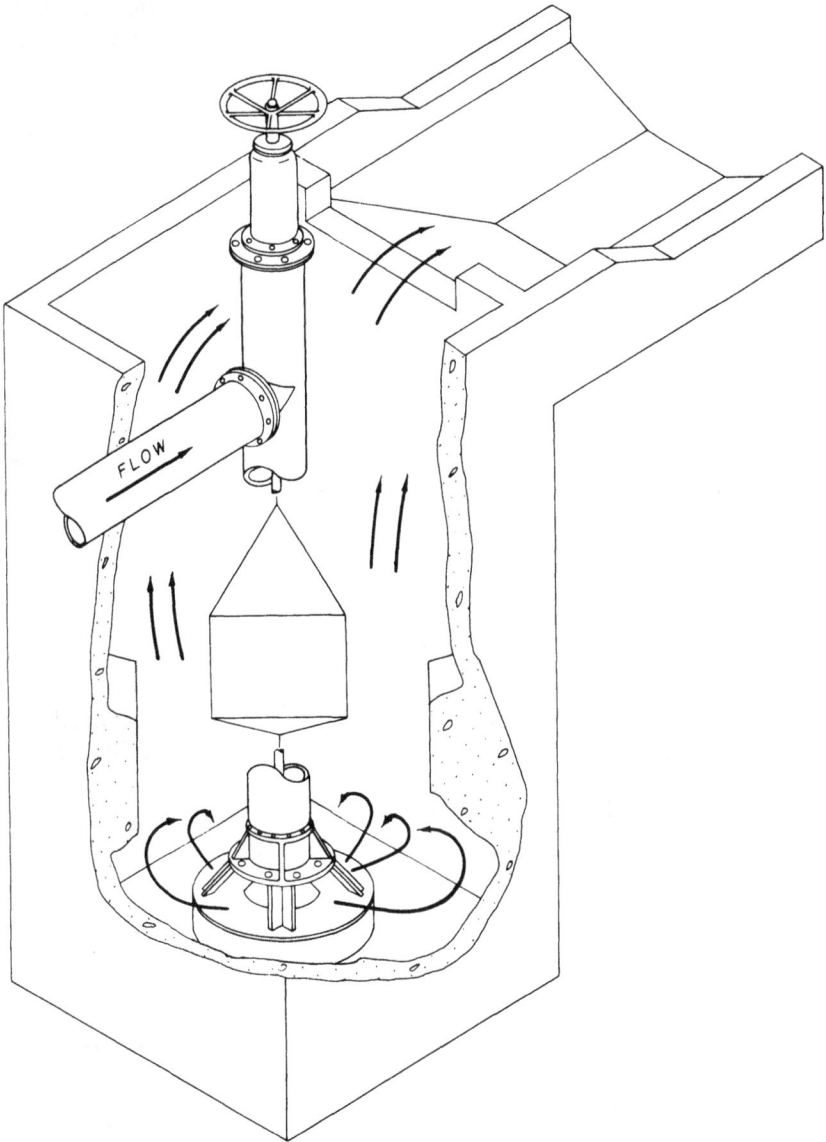

Figure 8-2.–Vertical stilling well. 103-D-1720.

features involved in siphon elbows that require special mention.

For a pumpline discharging into a canal, the transition from conduit to canal is accomplished partially in the siphon elbow. The diverging transition of the elbow influences both the priming action and the efficiency. Performance of the transition should be checked carefully in a model study. Siphon priming and air evacuation from the crown should be investigated during model operation. Design and operation of a model would be based for the most part on the Froude similitude. However, viscous flow may cause the siphonic action to be interrupted by an accumulation of air separating from the water in the low-pressure region. Part of the elbow near the crown may be modeled and operated near prototype pressures and velocities to study the viscous effect. A study of this type was made for the siphon elbows of the Grand Coulee Pumping Plant discharge lines [35, 36].

8-9. **Power Penstocks.**–Power penstocks are designed to conserve head for power generation. Thus, losses within the system are held to an economical minimum by using a minimum length of penstock, a large-diameter penstock, or by streamlining parts of the penstock. The length of the penstock is governed by the location of the powerplant in relation to the dam and is fixed before the problem reaches the laboratory. Also, in most laboratory studies, the penstock diameter has been previously determined because the most economical size can be computed by the designer. The laboratory study, therefore, involves streamlining parts of the penstock to obtain an optimum shape and may include investigating the relative merits of two or more proposed designs.

The trend toward using hydroelectric powerplants primarily to produce peaking power has made it economically desirable to build larger individual turbine-generator units. These units require larger bellmouth entrances, gates, penstocks, and auxiliary equipment such as hoists, trashracks, and many other associated items. Consequently, even a moderate reduction in the entrance bellmouth size could result in considerable savings in construction costs.

Studies consider entrance shape, conduit alinement, type and abruptness of bends, influence of branches, rate of change of area for expanding or converging sections, vortex formation, and the influence of obstructions, such as the trashrack structure over the entrance, the wheel- or roller-mounted gate frame at the entrance, or a butterfly valve within the line.

(a) *Entrance Shape.*–Design velocities in penstocks have increased in recent years and now approach 10 meters per second; thus, the shape of the entrance is important, but is still not as critical as one for an outlet conduit. The entrance should be flared to decrease the loss to an economical minimum; however, the size of the entrance opening should be as small as practicable to minimize the cost of the closure or bulkhead gate. The entrance cross section may be circular or rectangular, or it may be square with a transition to a circular conduit. Pressure taps or piezometers are installed in the model within and downstream from the entrance to permit measurement of the pressure gradient resulting from various shapes or degrees of streamlining. Abrupt changes in the pressure gradient are indicative of irregularities in the flow pattern and excessive head loss. A gradual change in pressure gradient is desirable. Optimum design is a compromise between good hydraulic efficiency and practical design and construction considerations.

(b) *Trashrack Loss.*–Trashrack structures stop large waterlogged pieces of wood and other foreign material from entering penstocks and damaging turbine runners. Trashracks are also used in outlet works to prevent clogging of gates and valves. Laboratory studies may be conducted to determine losses for special trashrack structures. Local concentrations of high-flow velocities should be investigated for forces that might induce vibration of trashrack members or cause excessive uplift or other adverse flow conditions.

Trashracks influence the pressure distribution within the penstock entrance. Of particular concern is the nearness of the bottom support of the trashrack to the penstock opening. The location becomes increasingly important as the velocities through the conduit increase, more so in outlet conduits than in power penstocks. Losses caused by obstructions such as gate slots, frames, etc., near the entrance can be minimized by laboratory investigation.

(c) *Emergency Control.*–If an emergency shutoff device, such as a butterfly valve, is to be placed in the penstock, knowledge of the head loss through the valve for normal operation is important. Piezometers and their locations play important roles in measuring the loss. Precise scale models should be studied at Reynolds numbers approaching that of the full-size structure.

(d) *Multiple Use of Penstocks.*–A penstock may be used as a temporary diversion tunnel during construction, operating as an open channel. The hydraulics of this usage are not complicated unless there are expanding or contracting sections involved, or if there is a large elevation difference between the exit and the entrance. In such cases, design of the penstock sections for flow control should receive special attention. A gradual transition that is satisfactory for closed-conduit flow may produce separation for the open-channel condition. The pressure gradient through the system should be investigated if the downstream end of the conduit is substantially lower than the entrance. If necessary, the gradient can be raised by placing a temporary cone or restriction at the downstream end of the conduit.

Use of reversible pump-turbine units for pumped-storage power generation places stringent requirements on the penstock entrance because it also functions as an exit during the pumping cycle. Velocity head recovery should be high when the structure acts as an exit. In turn, the head loss of the entrance should be low for the generating cycle. The entrance-exit structure thus becomes a special study for this type of plant. Careful use must be made of pressure and velocity-sensing equipment in finding flow characteristics of these structures.

(e) *Waterhammer.*–Closing or opening a gate or valve too rapidly induces dangerous pressure surges, possibly resulting in vibration and waterhammer–two important factors in studying transient phenomena in penstocks and pipelines. Studies by analytical or graphical methods are usually adequate for investigating these factors, with model studies being required only in special cases. Model studies involve application of the Mach similitude and use of special electronic equipment. Waterhammer problems are diversified, and a complete discussion of them in this text is impractical.

8–10. **Outlet Conduits.**–An outlet conduit releases water from a reservoir through or around the dam to the river downstream. An outlet includes several parts: trashrack structure; an entrance; main conduit; gate, valve, or other control device (which can be either at the entrance, in the line, or at the end of the conduit); and an exit from which the water discharges into an energy dissipator.

High efficiency of the conduit is important to minimize construction costs, but protection of the conduit from problems brought about by high velocities is of primary importance. Streamlining is necessary because the high velocities common in outlet conduits induce cavitation at irregularities or discontinuities in the flow boundary.

(a) *Streamlining.*–The reasons for streamlining outlet conduits and penstocks differ: for penstocks, the important consideration is minimizing head loss; for conduits, avoiding cavitation is of primary importance.

(b) *Trashrack.*–Trashracks can affect pressure distribution in the entrance to a conduit. Investigation of the relative positions of the bottom of the trashrack and the bottom edge of the outlet may be necessary.

(c) *Influence of Gate Slots.*–Gate slots or obstructions on the face of the dam near the entrance to the conduit or within the conduit may change the flow pattern sufficiently to induce severe subatmospheric pressures. Possibilities of cavitation should be investigated thoroughly and eliminated in the model studies.

(d) *Entrance Shape.*–The shape of the entrance should control the pressure distribution along the surface boundary to no lower than near-atmospheric pressures. Standard entrance shapes have been developed by general model studies for circular outlets normal to the face of the dam. One such shape (a type of "bellmouth") is the surface generated by revolving a quadrant of an ellipse about the axis of the circular outlet. The length of the semimajor axis of the ellipse quadrant is 0.5 of the outlet diameter, and the length of the semiminor axis is 0.15 of the outlet diameter. The position of the elliptical quadrant is such that the ellipse, at the end of its semimajor axis, is tangent to the face of the dam and, at the end of its semiminor axis, is tangent to the periphery of the circular outlet. If the entrance is not one of the standard shapes, the pressure distribution should be thoroughly investigated at the highest velocity available in the laboratory. Although some work has been done to establish standard shapes for square and rectangular entrances [37], model studies should still be made of such designs to determine the pressure distribution near the corners.

A bellmouth entrance designed for zero pressure will be subjected to subatmospheric pressures if the exit of the conduit is placed low with respect to the entrance and the conduit is not constricted near the downstream end. If the outlet pressures are lowered to the point that vapor pressure is approached at the entrance, there will be cavitation and, thus, severe pitting (cavitation erosion).

(e) *Conduit Alinement.*–Pressure reductions associated with abrupt changes in boundary alinement or short-radius curvatures should be investigated to ensure that there are no negative pressures that might induce cavitation erosion in the prototype.

(f) *Pipe Roughness.*–Model studies of conduit roughness are seldom required; the model flow surface is simply made as smooth as possible. It is sometimes necessary to shorten the model conduit to reduce the head loss and to represent computed pressure conditions of the prototype system. Scales producing Reynolds numbers of 1×10^6 or higher in closed-conduit models minimize the effect of viscous losses. Computations [29] are usually satisfactory for determining head losses caused by friction.

(g) *Conduit Exit Shape.*–The design of the exit end of an outlet conduit is governed by the required conditions in the area where the water is to be released from the system. If minimal spray is desired and the conduit terminates at the downstream face of the dam high above the tailwater surface, a special bend may be installed near the downstream face, together with a trough to direct the water down the spillway surface (see fig. 8-3). The end of the full-flowing conduit is usually in the form of a converging cone to offset the pressure reduction caused by the lower elevation at the exit. Conduit shaping at this location may include investigating methods of ventilating the boundary to prevent cavitation.

The shape of the trough should be designed to prevent large fins of water at the walls. An effective trough shape for a circular conduit exit is one with parallel sides spaced one exit diameter apart. The trough has a semicircular invert with a radius equal to that of the exit opening. This design is particularly appropriate for outlets not serving as regulators, that is, those which are either flowing full or have no flow at all. For conduits having a semicircular crown, vertical sidewalls, and a flat bottom, and operating with a free water surface, a trough having verical sidewalls and a flat bottom is more desirable. This shape, which facilitates aeration of the vertical curve to reduce subatmospheric pressures, is particularly suitable for conduits flowing partly full as required for regulation of the discharge.

Pressure intensity immediately downstream from the intersection of the trough and the face of the dam is of concern. The pressures should be near atmospheric to reduce the possibility of cavitation and instability in the flowing jet.

(h) *Effect of Spillway Flow on Conduit Exit.*–Water from an overflow spillway may pass over the exits of outlets terminating in the face of the dam. If so, study of pressure conditions and flow action at the conduit exit is essential whether the spillway and outlets are to operate either simultaneously or singly. The action of the spillway flow at the exit is of particular concern when water is not being released through the outlet. Spillway flow should not be allowed to impinge within the trough of the outlet, for such action produces considerable spray and can result in pressures causing cavitation damage.

(i) *Deflectors (Eyebrows) Over Outlet Trough.*–The spillway and river outlets shown in figure 8-3 generally do not operate simultaneously. Operation of the outlets, used to supply riverflow, is not normally required when the spillway must be used

A. Spillway flow passing over outlet deflectors. P801-D-79215

B. Deflectors (eyebrows) at upper ends of outlet exit troughs. P801-D-79214

Figure 8-3.–Grand Coulee Dam spillway and outlets.

141

to discharge larger quantities of water during flooding or other periods of large demand.

Aeration of the outlet conduit is provided by large vents placed immediately downstream from the control. Thus, when the spillway and outlets do operate simultaneously, the moderate atmospheric pressure in the conduit will not draw the trajectory of the spillway stream down appreciably as it passes over the upper edge of the outlet trough. However, aeration alone will not eliminate impingement of the spillway stream in the trough. A flow deflector on the face of the spillway upstream from the trough will raise the trajectory of the flow to clear the opening and prevent impingement in the trough (fig. 8-3B). Deflector shapes may be worked out in a model.

CHAPTER 9

Control of Flow

A. GENERAL

9-1. Introduction.–A control is defined as a device, usually a gate or valve, used to restrict or stop flow in a conduit or other water passageway. For this discussion, controls will be classified in four general groups in terms of the movements of their control elements:

(1) Those in which the flow is restricted by moving a leaf across the passageway, such as wheel-mounted and jet-flow gates;

(2) Those having a needle or plunger moving in the direction of flow to contact a seat, such as needle and hollow-jet valves;

(3) Those in which a rotating member permits free passage in one position and stops the flow when rotated 90°, such as plug, sphere, and butterfly valves; and

(4) Those which act to prevent backflow in a line, such as flap or check valves.

Flow through a conduit may be controlled by devices located at the entrance, within the conduit, or at the downstream end of the system. The devices may be adjustable or nonadjustable, depending on the outlet capacity and requirements downstream. Adequate aeration immediately downstream from the control is essential in preventing cavitation and vibration, particularly for a control used for regulation of high-velocity flow. Aeration is less important for a control operating only in the open or closed positions; however, sufficient air should be provided to minimize vibration during the opening and closing cycles. Pressure changes within the conduit system caused by the control device require thorough investigation.

9-2. Control Requirements.–Basic design problems are similar in all types of control gates or valves. They must be built to withstand the design pressures. The operating mechanism must overcome frictional drag or unbalanced hydraulic forces occurring during the opening and closing movements. Discharge capacity must be sufficient to satisfy design requirements and yet the gate or valve must be of economical construction. The control must be able to operate through the required range of heads at various openings with a minimum of vibration and without cavitation damage, either to the device itself or to the downstream conduit. Also, adequate seals must be provided to assure a minimum of leakage.

9-3. **Laboratory Tests.**–In the development of large valves or gates for high-head installation where there is no precedent for design, the study of the hydraulic characteristics may include laboratory tests on scale models. The studies include:

- Evaluating the unbalanced forces occurring on moving parts during the operating cycle,
- Measuring discharge capacities,
- Recording pressure intensities, paying particular attention to severe subatmospheric conditions that might induce cavitation,
- Reshaping questionable flow passageways to prevent cavitation,
- Providing adequate air relief in critical low-pressure areas,
- Measuring torque of rotating elements, and
- Studying the adequacy of seals.

A gate or valve operating at partial openings dissipates a large amount of energy; if vibration occurs, studies should include the contributing variables. Gates and valves and the conduits they are installed in are scaled according to Froude relationships. The scale of the model is selected to place important flows well within the turbulent range to minimize viscous effects.

9-4. **Force and Torque Measurements.**–Forces in hydraulic model testing are measured both directly and indirectly. The direct method is that of weighing or using a calibrated force-recording sensor. There are many such sensors and only the principles of measurement will be mentioned here. Displacement of a spring or lever; elongation of a metal bar, rod, or wire; or torsional displacement of a metal rod or other structural shape may be used to indicate forces directly. The indirect method computes the force by integrating the pressure distribution curves for the surface or portion of the structure under study. Both methods are useful in evaluating the forces acting on the moving portions of control devices and the torque to be overcome in the operation of these devices.

Force or torque on an operating mechanism or rotating part is usually measured by weighing. Equipment may consist of a lever or system of levers coupled to a commercial scale or specially designed apparatus. The arrangement should be such that the frictional resistance in the model can be evaluated or eliminated from the force or torque measurements.

9-5. Leakage.–Leakage past control devices normally increases with time and can cause problems in the field when making inspections, performing routine maintenance, or making needed repairs. To minimize leakage, good seal design and close dimensional control are necessary. Models operated according to Froude's law are inadequate for studying seals; instead, a special test facility that will accommodate a section of the prototype seal and can be operated at prototype heads must be used. Seals for gates and valves are discussed in detail at the end of this chapter.

9-6. Discharge Capacity.–Devices for regulating and controlling flow in closed conduits are compared in terms of their capacity to discharge fluids under a given head. Head/discharge curves are useful in making such comparisons. These curves can be computed from the relationship between coefficient of discharge, gate opening, and head. The magnitude of the discharge coefficient depends on the flow conditions and on the area used in the calculations. The area should always be defined when presenting coefficient data. The area used may be that of the upstream conduit, the gated flow passage, the clear opening of the throttled section, or the downstream pipe diameter. Thus, the indicated discharge coefficient, C, computed from the relationship $Q = CA\sqrt{2gH}$ or $Q = CA\sqrt{2g\Delta H}$ might vary widely for controls having the same capacity, Q, depending on the choice of area, A. Head, H or ΔH, may be a pressure head value or the total head containing both velocity head and static pressure head. As with area, the head value must be clearly defined. Included in the definition must be the location of the head-measuring station (or stations if a differential value is used).

9-7. Air Requirements.–Air admission will normally be required in the conduit downstream from a control to minimize subatmospheric pressures and possible cavitation damage. Investigations of the air-vent size should consider:

145

- The limiting subatmospheric pressure that can be tolerated in the structure,
- The economics of constructing either one large or several small vents,
- Maximum allowable air velocity within the duct system,
- Maximum allowable air velocity at the entrance to the duct, and
- Effect of the quantity of air on flow characteristics of the water in the downstream conduit.

Prototype correlation with model air demand measurements is normally scaled by Froude relationships. Success, however, has been limited. In recent years, analysis indicates that proper techniques have not been used in modeling the air ducts. Detailed attention has been given to the geometric scaling of the waterflow passages of the model; however, little attention has been paid to modeling an air vent to produce the prototype velocity profile of the airflow above and around the water jet. Analysis shows that the velocity profile can vary depending on air inlet position, pressure gradient in the conduit downstream from the control, jet roughness, and the pressure of the air at the supply point [38].

Models of air vents that include only an air-measuring device (usually an orifice) and a short length of pipe do not necessarily model the air pressure and quantity, nor form the correct velocity distribution. Scaling an air-vent system to the size of the model or scaling the head loss between inlet and outlet of the prototype vent will produce a better prediction of prototype air demand. In the laboratory, modeling the complete air-vent system may not be possible or desirable. Instead, the introduction of a head loss in the model duct system representing the ducting loss of the prototype is the most practical way to scale the air system. Air introduced at the proper pressure and outlet location in the model should produce an acceptable air velocity profile and, therefore, more closely indicate the air quantities of the prototype.

The size of the vent entrance and duct should limit the air velocity in the duct to a maximum of 100 meters per second. This speed is associated with a whistling sound and possibly an objectionable noise level. Air speeds of about 20 meters per second near the duct entrance will normally not endanger people and will limit the number of loose objects swept into vents. Vents should not contain abrupt bends and sharp corners that would induce local high velocities and noise levels. Both mathematical and hydraulic models are used to establish the quantity of air required to ventilate an outlet conduit and the magnitude of the pressures within the conduit.

B. CLASSIFICATION OF CONTROLS

9-8. Introduction.–Although all outlet controls serve a similar purpose, their mechanical and hydraulic characteristics may differ widely [39]. For this reason, brief descriptions of the various types of controls and their outstanding characteristics and peculiarities are presented here.

9-9. Bulkhead Gates.–A bulkhead gate isolates or separates a portion of a system or structure from its other parts. One common type is used to unwater conduits that have their entrances on the upstream face of the dam. This gate is usually lowered by crane down the face of the dam to cover and seal the conduit entrance under balanced-pressure, no-flow conditions. The hydraulic problems concerning bulkhead gates operating under balanced conditions are few. Model studies are made only when water in motion influences the placing or operation of the bulkhead. These studies concern positioning of the gate and measuring line pull or required hoisting capacity. When transportation by floating in water is required to place the gate or special bulkhead in position, studies include flotation and stability problems and the number, size, and positions of the lines required to maneuver the gate into position. Studies of this type are discussed in section 11-14, Floating Equipment.

9-10. Roller- and Wheel-Mounted Gates.–

(a) *General.*–Like bulkheads, roller-mounted and wheel-mounted gates, also known as "tractor" or "coaster" and "caterpillar" or "fixed-wheel" gates, respectively, are used to unwater conduits under balanced-head conditions for inspection and maintenance purposes. In addition, they are capable of closing by gravity under unbalanced head, that is, with water flowing through the conduit; most bulkheads and other slide-type gates do not have this capability because of excessive frictional forces. Wheel-mounted gates have, on occasion, also been used as regulating gates on spillways.

Roller- and wheel-mounted gates can be installed at the conduit entrance or within the conduit. Tracks and guides are provided to maintain alinement of the gate. Hydraulic hoists are normally used for operating these gates. The major problems encountered with roller- and wheel-mounted gates are vibration caused by hydraulic conditions, seal damage, and the downpull and uplift forces that develop during unbalanced closure [40].

(b) *Hydraulic Downpull.*–When a roller- or wheel-mounted gate is lowered across a conduit that has water flowing through it, the water velocity past the surfaces of the gate leaf varies with the position of the gate and the nearness of each surface to the flow opening. The higher the velocities are, the lower the static pressures on the surfaces of the gate will be. High velocities beneath the gate cause the static pressures on the bottom surface to be low relative to those on the top; thus, an unbalanced downward force, known as hydraulic downpull, will exist because of these pressure differentials. Because the velocity through the conduit at the time of closure has a major influence on the magnitude of the downpull force, the efficiency of the conduit control downstream is a factor to be considered and evaluated in hydraulic model tests. The factors influencing the downpull force, discussed in the following paragraphs, affect the design and capacity of the operating hoist:

- Entrance elevation–The depth of the conduit below the reservoir water surface is usually established by the physical features of the dam and appurtenances. Therefore, the maximum head under which the gate must operate is not a variable in most hydraulic investigations.

- Gate width and thickness–For a particular opening and shape of the bottom surface, the minimum downpull will result when the thickness of the gate is a minimum and the width of the gate is made nearly equal to the width of the opening. This ratio of gate width to opening width is usually fixed by structural requirements and is not determined by hydraulic investigation. The effect of gate thickness on the downpull force for a particular design is shown on figure 9-1.

- Bottom surface shape–The contour of the bottom surface of the gate influences the velocity and pressure distribution: the lower the velocity, the higher the static pressures, and the smaller the downpull force. Thus, the bottom surface shape that produces the lowest velocities along the surface is the most effective in reducing downpull (fig. 9-2). If the lip of a gate is located close to the face of the dam, the area

148

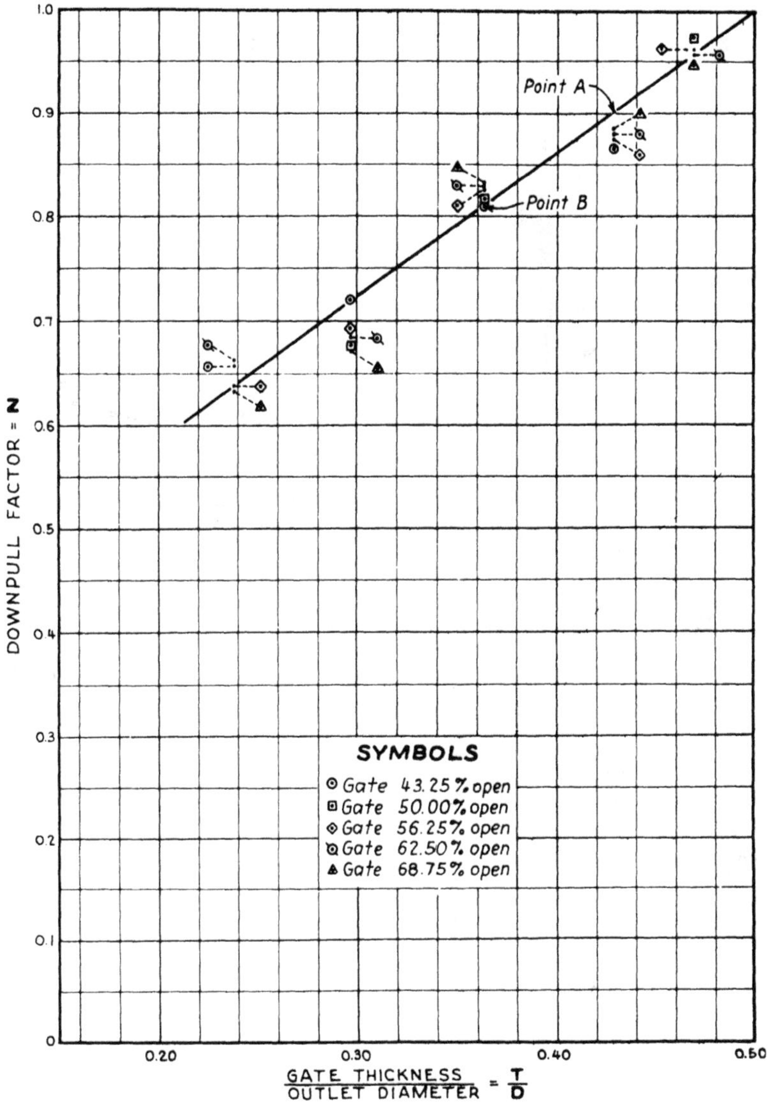
DOWNPULL FACTOR = N
1.0
0.9
0.8
0.7
0.6
0.5
0.4
0.3
0.2
0.1
0
0.20
0.30
0.40
0.50
GATE THICKNESS / OUTLET DIAMETER = T/D
Point A
Point B
SYMBOLS
Gate 43.25% open
Gate 50.00% open
Gate 56.25% open
Gate 62.50% open
Gate 68.75% open

NOTES

Tests were made by varying thickness of final design gate on Shasta Dam Outlet model.
The change in downpull force, F, by a change in gate thickness from T_1 to T_2 may be expressed as:

$$\frac{F_1}{F_2} = \frac{N_{\frac{T_1}{D}}}{N_{\frac{T_2}{D}}}$$

Figure 9-1.–Effect of gate thickness on downpull. 103-D-1721.

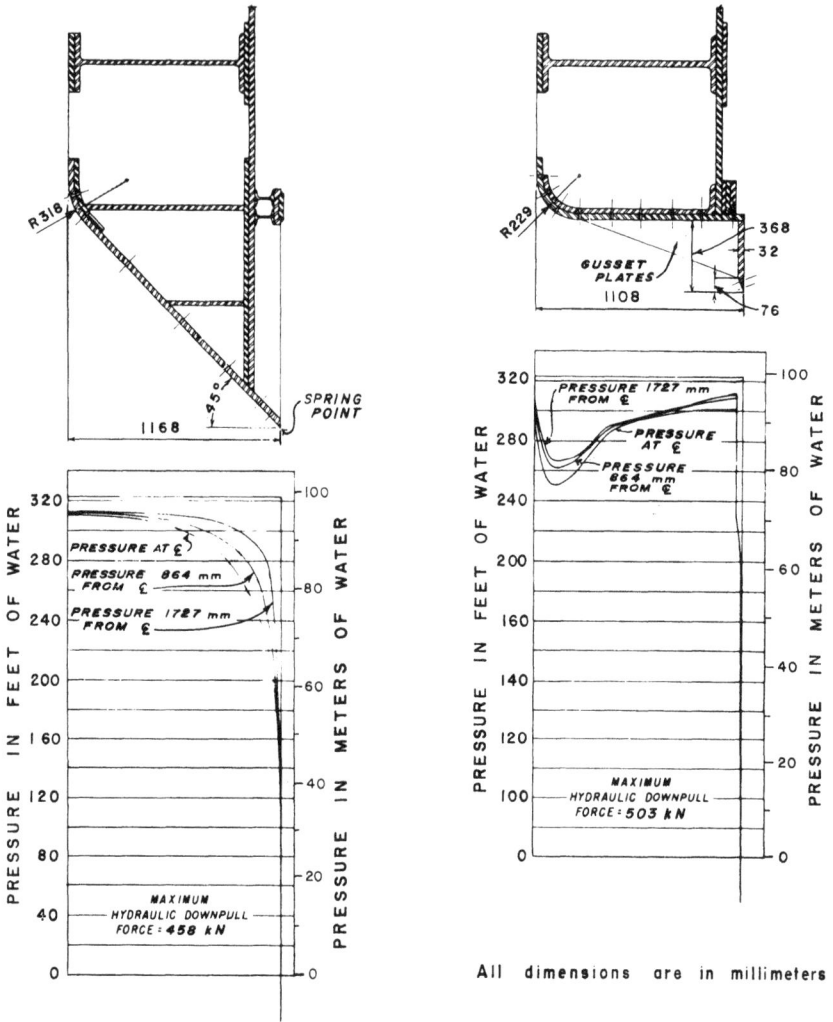

Figure 9-2.–Effect of gate bottom shape on pressure curves as determined by model studies. 103-D-1722.

available for the pressure acting in an upward direction is increased and may result in a lifting force on the gate.

● Seal–The location of the gate seal and the type of seal used may have considerable effect on the magnitude of the downpull force. Model study should include investigation of these effects.

- Trashrack–Pressure distribution in a conduit entrance is related to the nearness of the base of a trashrack structure to the opening. Thus, the trashrack base influences both the discharge capacity and the downpull force during unbalanced closure of the gate (fig. 9-3).

- Conduit pressure and gate recess–Pressure in the conduit immediately downstream from a gate decreases as the opening decreases, and can become subatmospheric. The magnitude of the subatmospheric pressure depends on the nature of the restriction or control in the conduit downstream. The effect of the downstream control on downpull is shown on figure 9-4.

Conduit pressure influences downpull in two ways. First, an increase or decrease in the net effective head at the gate opening causes changes in velocity and pressure differential, resulting in a change in downpull force on the gate itself. Second, variations in conduit pressure exerted on the lower surface of an extended top seal will vary the differential head on the seal, also influencing gate downpull. Air admission by venting downstream from the gate reduces the downpull on the seal by increasing the static pressure on the low-pressure side of the seal. A further reduction in downpull can be gained by placing a recess in the face of the dam (fig. 9-5). This recess equalizes the pressures on the bottom and top sides of the top seal and thus reduces the downpull force contributed by the seal. Investigation of the shape of the recess may be included in the model study.

(c) *Downpull Measurements.*–The hydraulic downpull on a model may be determined by direct force measurements or by integration of pressure distribution curves[41]. Direct force determinations are usually made by providing a mechanical or electrical measuring instrument in the operating mechanism. Records should be obtained with the gate moving upward as well as downward to enable isolation of frictional resistance. The mass of the gate and the net downward force produced by the gate when suspended in water should be measured to enable determination of the true hydraulic downpull force. Piezometers are often installed in the gate to establish pressure distribution curves from which the hydraulic downpull force can be calculated by integration. Many piezometers are placed in the bottom surface, a few in the top surface, and others in areas judged by the investigator to be critical. Records should be obtained for several positions of the gate to avoid overlooking a critical condition.

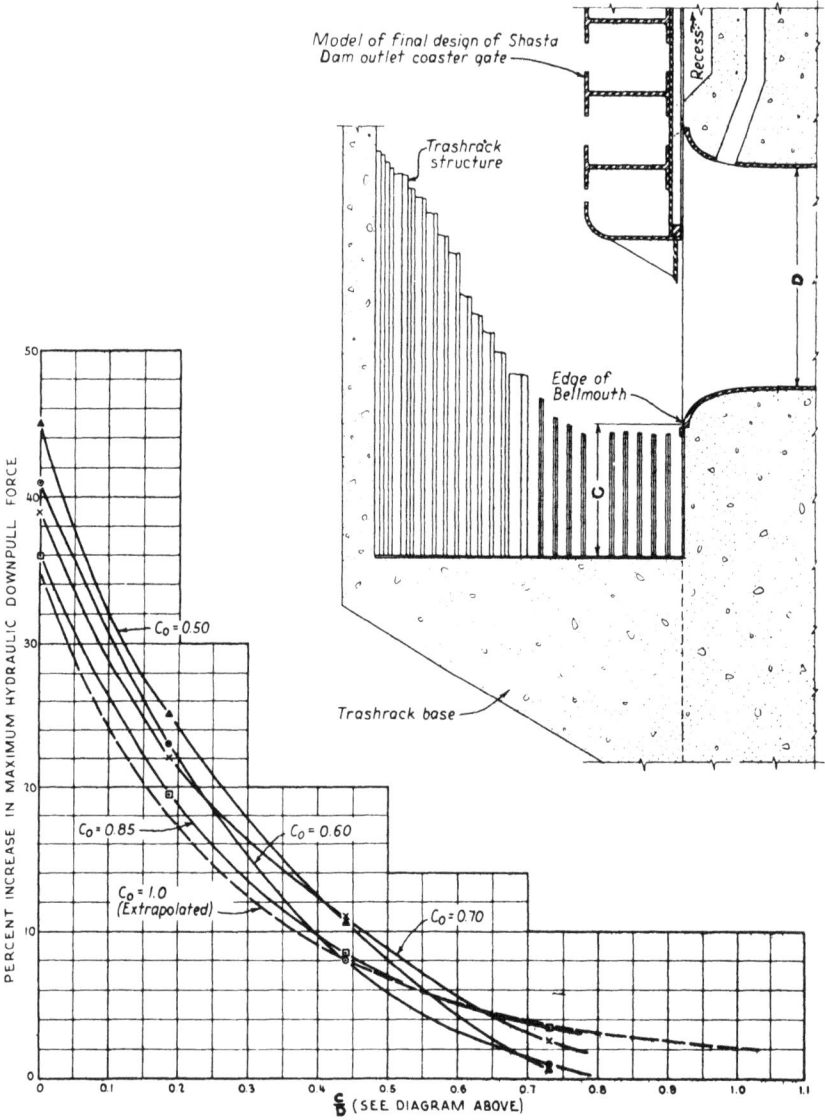

Figure 9-3.–Effect of trashrack base on hydraulic downpull. 103-D-1723.

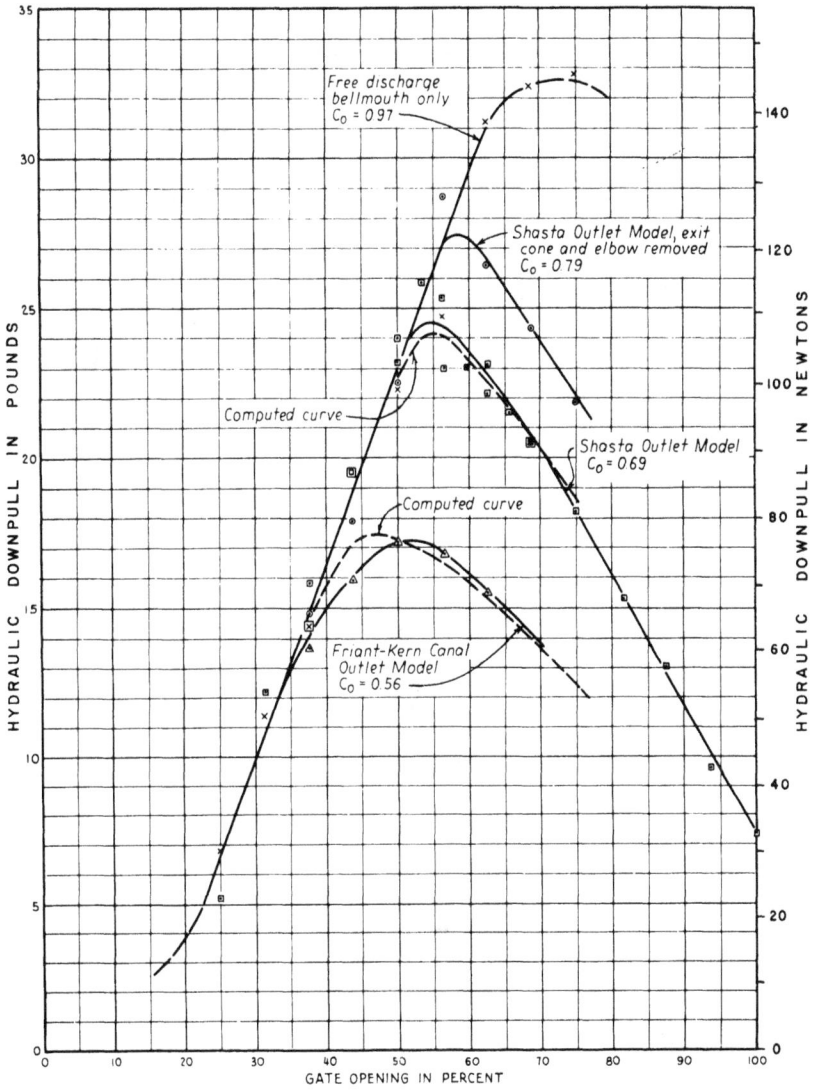

Figure 9-4.–Effect of restricting outlet flow on downpull. 103-D-1724.

153

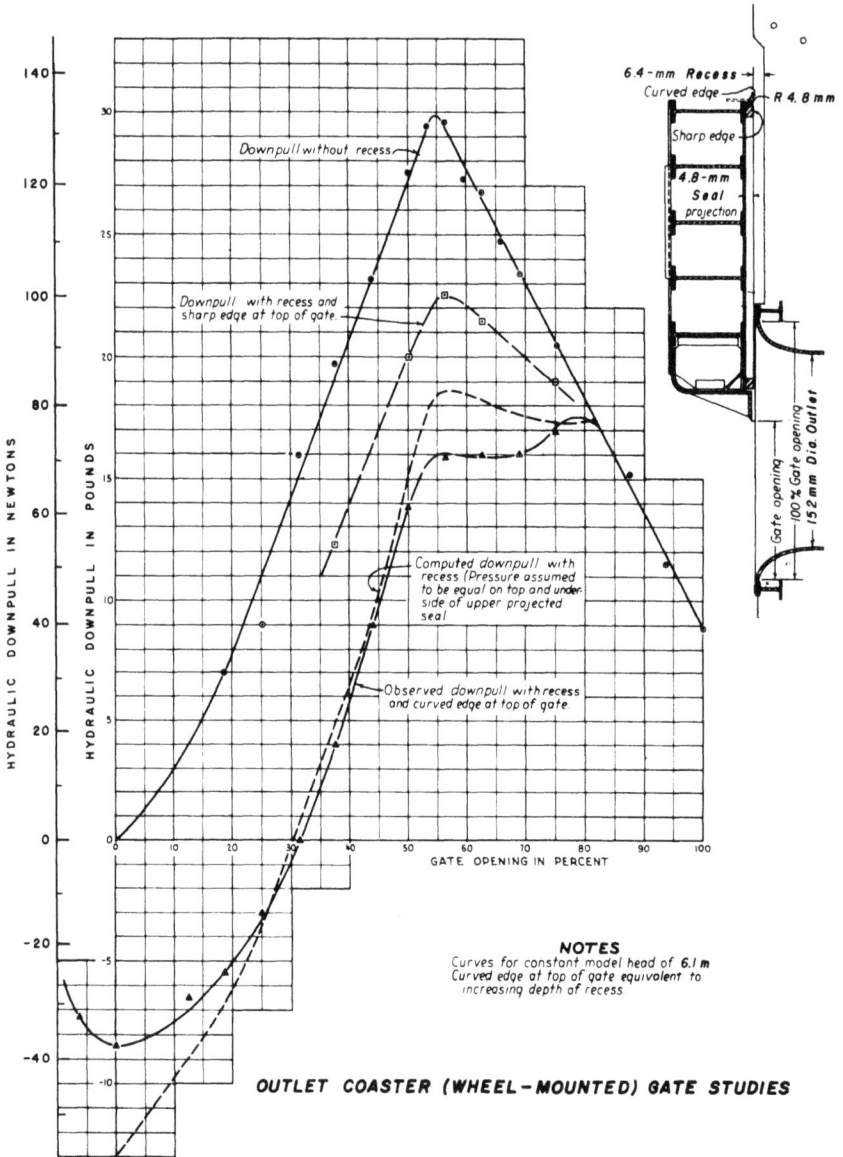

Figure 9-5.–Effect of recess on downpull. 103-D-1725.

9-11. **Slide Gates.**–Slide gates are used to control flow through conduit systems. Well-lubricated guides help reduce sliding friction; this is particularly important when the conduit is large and the head is high. Use of the slide gate for regulation necessitates aeration of the conduit immediately downstream and special parts or treatment in the housing or gate well, unless the entire system is under sufficient pressure to prevent development of vapor pressure.

(a) *Ring-Follower Gates.*–The most efficient outlet conduit is one having continuous flow boundaries and no obstructions. A ring-follower gate has been developed for use in circular conduits where regulation of discharge is not required and where high efficiency and smooth flow without subatmospheric pressures are required. The leaf of this gate moves perpendicular to the axis of the conduit. To shut off the flow when the gate is closed, the upper end of the leaf is solid; the lower end has an opening the same diameter as the conduit to form a smooth, continuous passageway when the gate is wide open. The name of the gate originates from the fact that the ring, or the end of the leaf with the opening in it, follows the shutoff portion of the leaf as the gate is opened. Ring-follower gates are not used for regulation, because of possible vibration and cavitation problems.

(b) *Regulating Leaf Gates.*–The leaf-type slide gate is adaptable for regulation of flow at high heads. In high-head installations, special consideration must be given to the shape and size of the gate slot and the pressure and flow conditions within it. Flow through the gate at partial openings is complicated in that the degree of contraction at the bottom edge of the gate is much greater than at the walls below it. The tendency is for the flow to be deflected downward and sideward at steep angles, impinging in the gate slots and producing undesirable pressures and wave conditions in the slots and housing. Wide-slot gates (fig. 9-6) are subject to cavitation erosion at high-operating heads. Narrowing the slot and reshaping the leaf provides a high-capacity gate for high-head installations (fig. 9-7). Sealing at the upstream side of the leaf reduces the hydraulic downpull and allows the gate leaf to be shaped to cause the flow to spring free of the gate. If the seal is on the downstream side and the pressure is relieved in the upper part of the housing or bonnet of the gate, an uplift instead of a downpull is possible, depending on the shape of the bottom of the leaf. Careful laboratory analysis of flow through high-pressure gates is necessary.

(c) *Jet-Flow Gates.*–The jet-flow gate, developed by the Bureau of Reclamation, is a relatively inexpensive regulating device

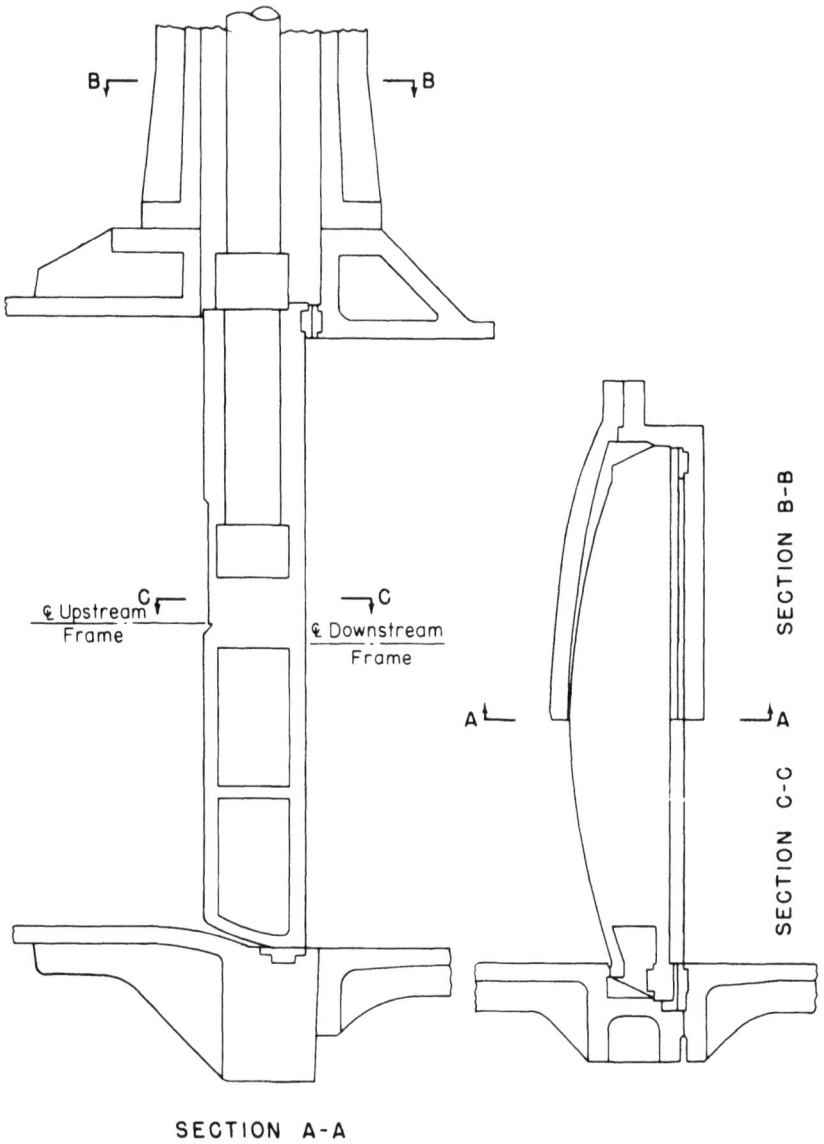

SECTION A-A

Figure 9-6.–Regulating slide gate (wide slot). 103-D-1726.

156

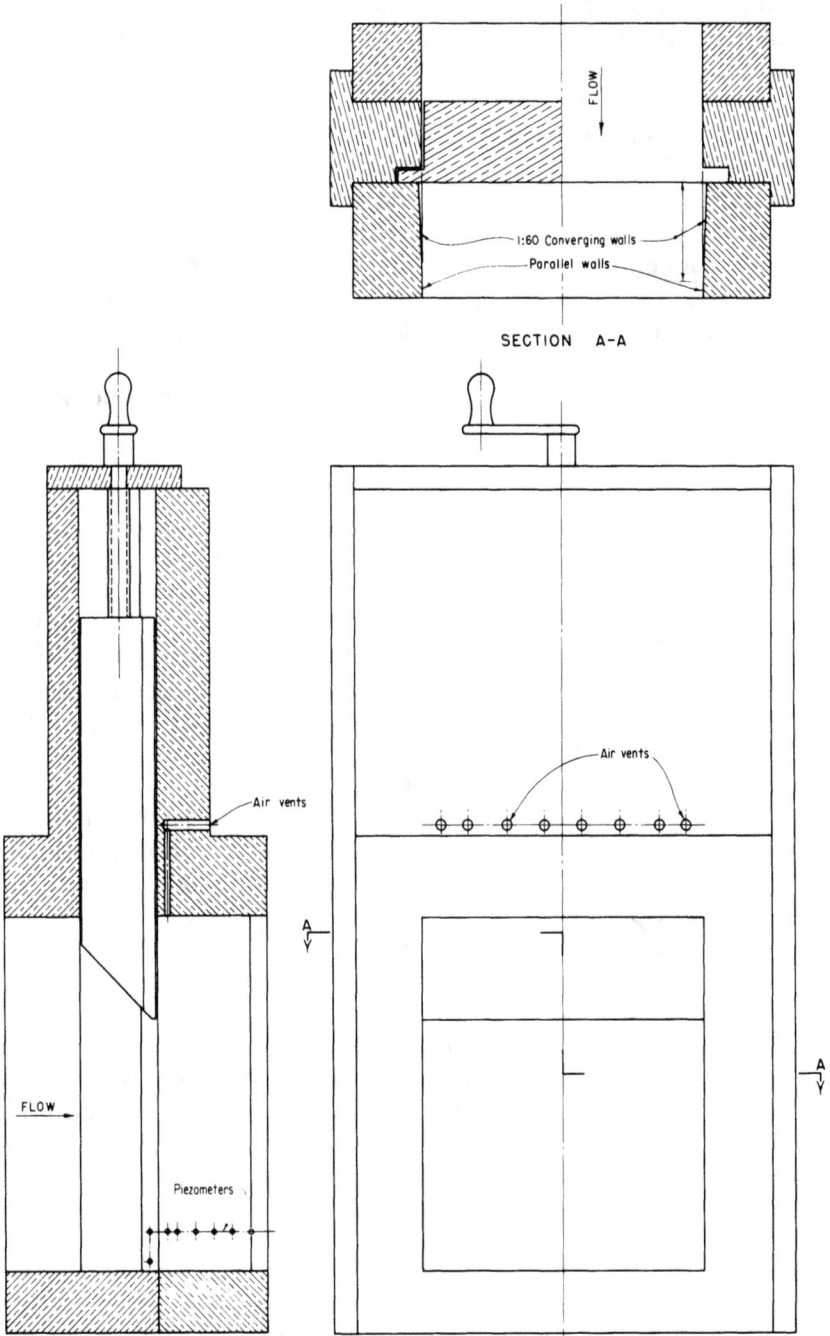

Figure 9-7.–Model of regulating slide gate (narrow slot). 103-D-1727.

for high-head outlet works. This gate consists of a movable gate leaf within a body having a circular orifice at the upstream side of the slot and a downstream conduit of a shape that will provide adequate aeration for the jet. The upstream face of the gate leaf is smooth and remains in contact with a seal contained in the orifice side of the gate body. Sealing at the upstream face eliminates the hydraulic downpull inherent in gates with downstream seals.

Contraction of the jet after it passes through the orifice prevents the jet from entering the gate slot, thereby eliminating the danger of cavitation downstream from the slot. The gate is of relatively simple construction and is capable of trouble-fee operation at any gate opening. Fully opened, the gate has a discharge coefficient of about 0.79, based on the total head in the approach conduit and the area of the orifice.

Three configurations of the upstream gate body are shown on figure 9-8. The configuration used at Shasta Dam for the first installation of the jet-flow gate included an expanding section formed by a simple curved surface tangent to the upstream conduit wall and intersecting the 45° orifice cone (fig. 9-8A). This shape, designed for use with a tube valve that was replaced by the jet-flow gate, was changed to a conical expanding section in a later design.

The conical expanding section (fig. 9-8B) was used for a gate installation at Trinity Dam. This shape is less expensive than that used at Shasta Dam and possesses almost identical flow characteristics and discharge coefficients.

A new type of small jet-flow gate, designed to be installed in a conduit without an expanding upstream section, is shown in figure 9-8C. Such a gate includes many of the beneficial flow characteristics of the other two designs having more sophisticated entrance shapes, and is relatively inexpensive to fabricate.

9-12. **Radial Gates.**–Radial gates are generally used to control open-channel flow, but they may also be used in closed conduits. The problems in closed conduits are similar to those experienced in open-channel flow; however, such studies should, in addition, include design of the seals, the forces acting on the gate, and the pressure distribution on the gate and in the conduit downstream. Small irregularities in water passage surfaces, unimportant in open-channel flow under low head, may induce cavitation in the closed conduit.

9-13, **Needle Valves.**–Several types of needle valves have been used as outlet controls [42]. Early valves were designed such that at small openings the surfaces of the valve body and the needle in

158

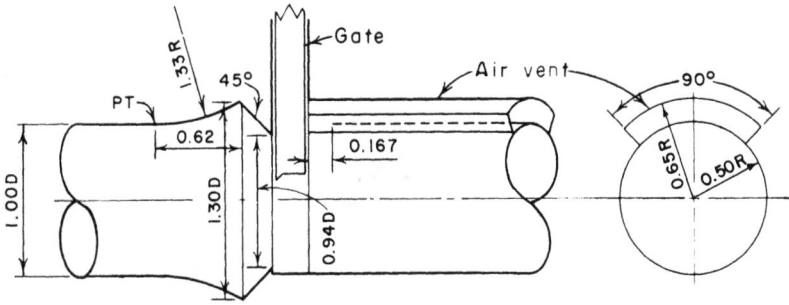

A. SHASTA DAM JET-FLOW GATE

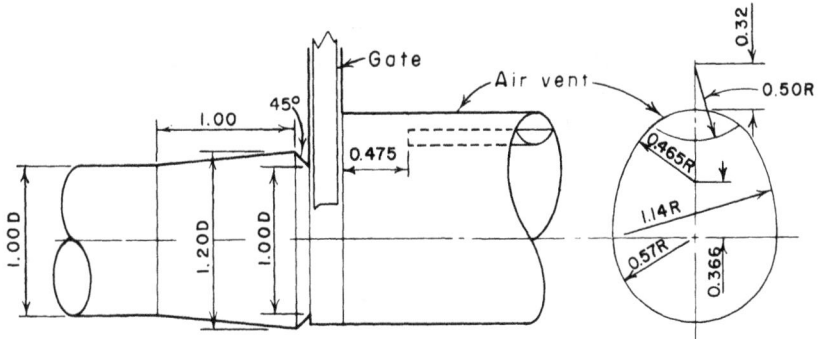

B. TRINITY DAM JET-FLOW GATE

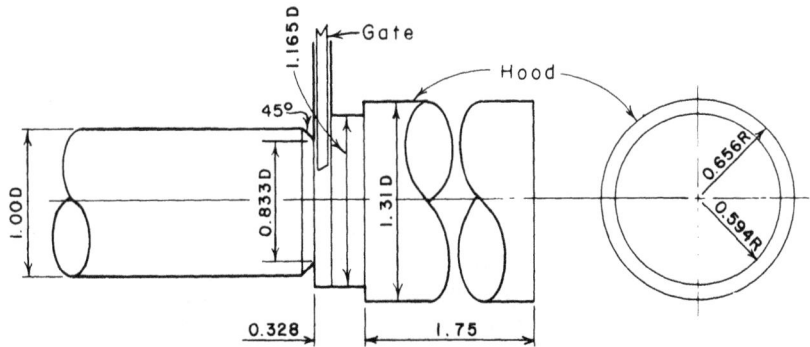

C. EAST CANYON DAM JET-FLOW GATE

Figure 9-8.–Jet-flow gate configurations. 103-D-1728.

the flow passageway were divergent. This divergence, which conformed to the constant-area flow-passage criterion at full valve opening, provided a high rate of discharge, but induced severe subatmospheric pressures when the valve was operated under high heads. Cavitation resulted and redesigning was necessary to produce converging surfaces in the flow passages. Also, a sharp spring point at the body exit was incorporated. (See fig. 9-9.) The jet of water emerging from a needle valve is very stable unless cavitation is present near the exit. Alternate increase and decrease of pressures near the cavitation zone may induce an unstable jet and cause fluttering and intermittent spray from the jet surface immediately downstream from the valve.

9–14. **Tube Valves.**–A tube valve is essentially a needle valve with the tip removed from the downstream needle (fig. 9-10). Hydraulic characteristics are similar to those for the needle valve. The jet is not so variable in size for changes in valve opening as compared to the needle valve, but it becomes unstable at small openings and tends to flutter and disintegrate into a cloud of spray. Electrical analogy studies are helpful in determining the initial degree of streamlining for the flow passage; however, care must be exercised in analyzing the data and attempting to carry the study beyond the limitations of the apparatus.

9–15. **Hollow-Jet Valves.**–The hollow-jet valve (fig. 9-11) resembles the upstream half of a needle valve. It evolved as a result of efforts to reduce weight and decrease the cost of a control for the exit end of a conduit. This valve is well suited for high-head applications since it does not tend to have cavitation problems except at valve openings of less than 5 percent. Deviations from the flow passage shape, shown on figure 9-11, can result in subatmospheric pressures, inducing cavitation. The hollow-jet valve has about 18.5 percent more capacity than the same size needle valve, and the body is not subjected to full reservoir head. The stream from the hollow-jet valve emerges as a hollow cylinder having a constant outside diameter, and the flow is distributed over a wider area than from the needle valve. Dissipation of energy is thus facilitated. Pressure conditions, capacity, and ease of operation are the main factors considered in model tests of hollow-jet valves.

9–16. **Butterfly Valves.**–The butterfly valve, used in circular conduits, consists basically of a circular leaf that rotates about a horizontal or vertical diametrical axis. The valve is used extensively for both no-flow closure and for emergency closures of power

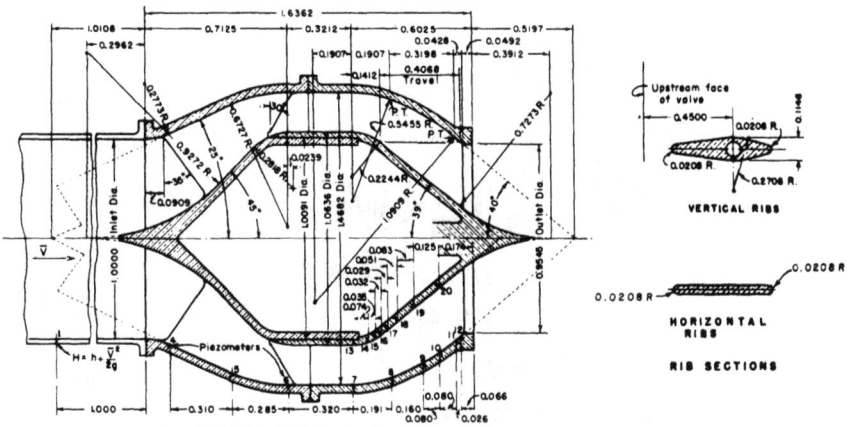

A. SECTION THRU VALVE
Note - All dimensions related to a unit inlet diameter

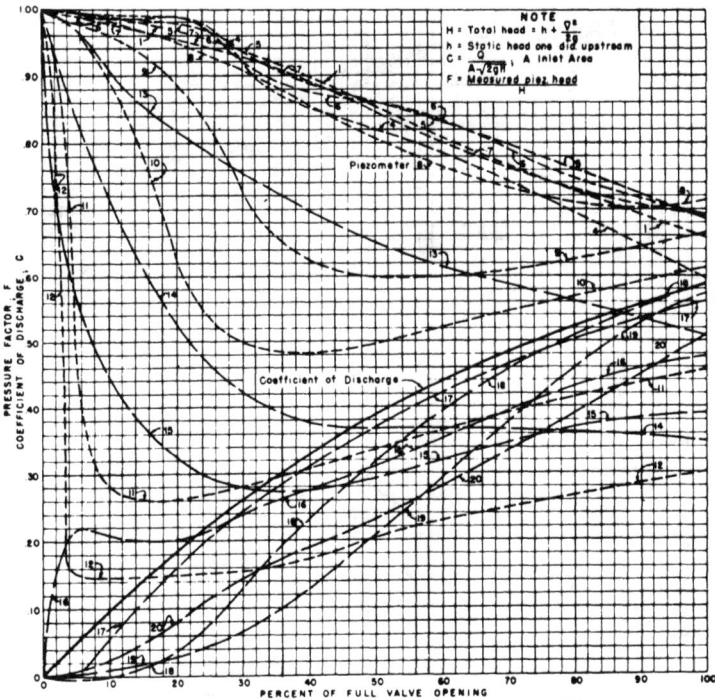

B. COEFFICIENT AND PRESSURE FACTOR CURVES

Figure 9-9.–Needle valve cross section and hydraulic characteristics based on model studies of a 150-millimeter (6-in) valve. 103-D-1729.

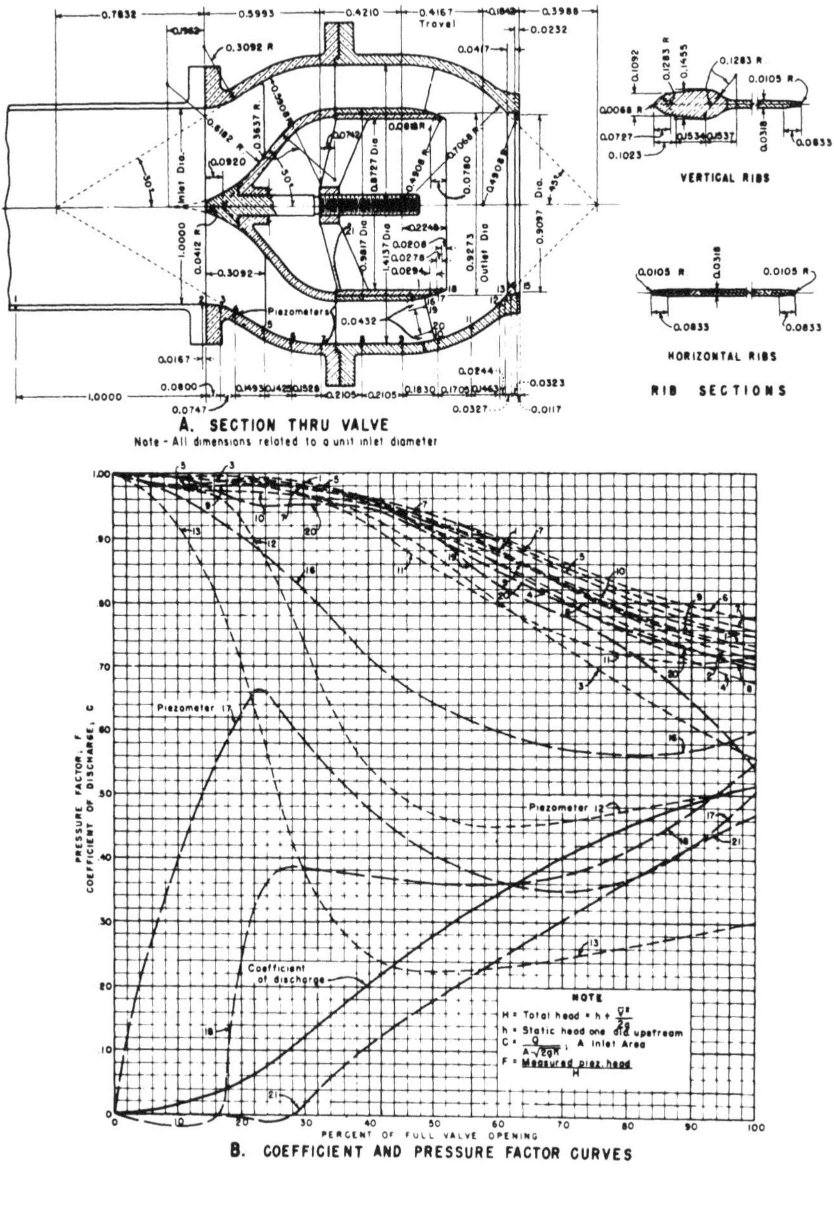

Figure 9-10.–Tube valve cross section and hydraulic characteristics based on model studies of a 150-millimeter (6-in) valve. 103-D-1730.

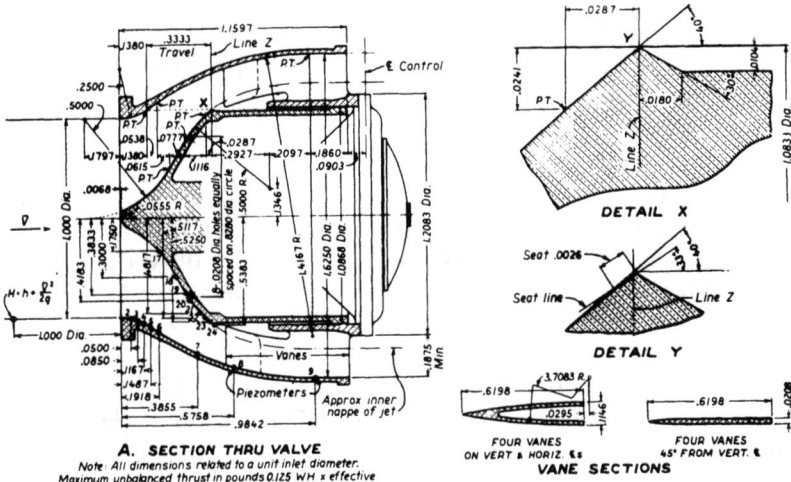

Figure 9-11.–Hollow-jet valve cross section and hydraulic characteristics based on model studies of a 150-millimeter (6-in) valve. 103-D-1731.

163

penstocks, and has occasionally served as a regulating device within or at the end of an outlet conduit. Cross-sectional shapes of the leaf and the valve body are very important for either application. When used for emergency closure, the valve operates either in the wide-open or the closed position. At partial openings, improper water passage shape can cause severe subatmospheric pressures and cavitation. Cavitation may also occur with high velocities when the valve is at or near the wide-open position. Studies of the valve should include the discharge coefficient, opening and closing torque, pressure distribution, jet configuration for free flow, and the sealing characteristics. Figure 9-12 shows a butterfly valve model under investigation. Free-discharge characteristics of a 150-millimeter butterfly valve are shown on figure 9-13.

9-17. **Horizontal Multijet Sleeve Valve.**–The horizontal multijet sleeve valve is an outgrowth of Bureau of Reclamation laboratory studies conducted on vertical stilling well and sleeve valve designs.

One application for sleeve valves is in long aqueducts for municipal and industrial water supplies. At present, control stations along aqueducts commonly utilize butterfly valves to regulate flow. Butterfly valves are limited to pressure-head differentials of

Figure 9-12.–Investigation of butterfly valve opening and closing characteristics. P801-D-79245

No extension downstream of valve

Open area normal to flow at wide-open position = 11 882 mm^2

Coefficient of discharge (wide-open valve) = 0.595

Discharge = 159 L/s at full open position with 7.09 m H$_2$O static pressure in line at valve entrance.

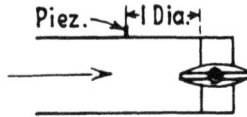

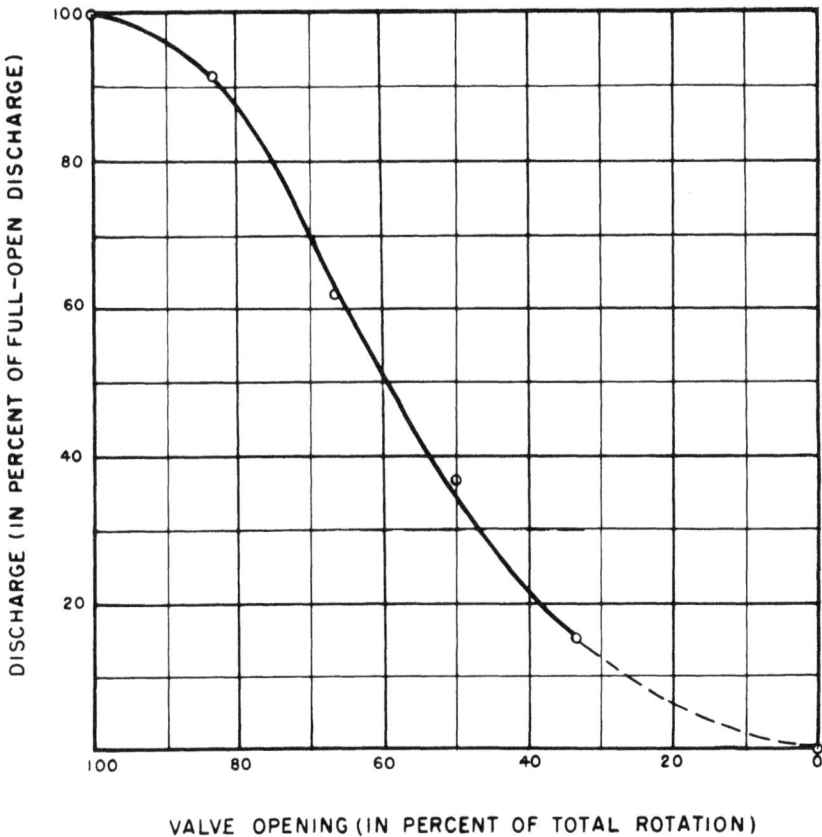

Figure 9-13.–Free-discharge characteristics of a 150-millimeter (6-in) butterfly valve. 103-D-1732.

approximately 15 meters, a restriction imposed to prevent cavitation damage to the valves. Thus, development of a valve and energy dissipator arrangement that could accommodate pressure differentials greater than 15 meters would be desirable to reduce the number of control stations required.

Horizontal multijet sleeve valves have been found to be compatible with aqueduct systems and they (1) adequately dissipate high-energy flow at small discharges and (2) pass design flows with a minimum energy loss.

The basic concept of the multijet valve and stilling chamber is illustrated on figure 9-14. Flow enters the valve from the left and is discharged through the perforated body of the valve into the stilling chamber. A cylindrical sleeve, located inside the valve, travels over the perforated section of the valve, controlling the port area, and thus, the valve discharge. Flow enters the downstream pipeline at the lower right of the stilling chamber. Figure 9-15 illustrates characteristics of an ideal multijet sleeve valve tested in the laboratory.

The use of horizontal multijet sleeve valves in municipal and industrial water supply aqueducts will save hundreds of thousands of dollars for each large flow-control station of conventional design that is eliminated. The first horizontal multijet sleeve valve was installed in 1977 on the Bureau's Frederick Aqueduct, Mountain Park Project, Oklahoma.

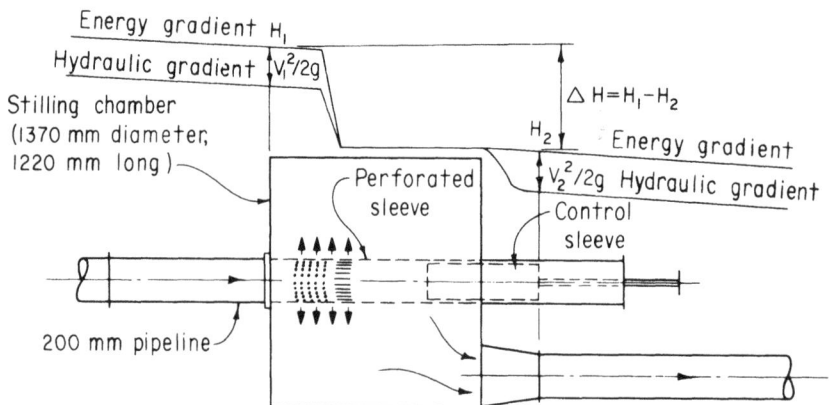

Figure 9-14.–Basic concept of horizontal multijet sleeve valve and stilling chamber. 103-D-1733.

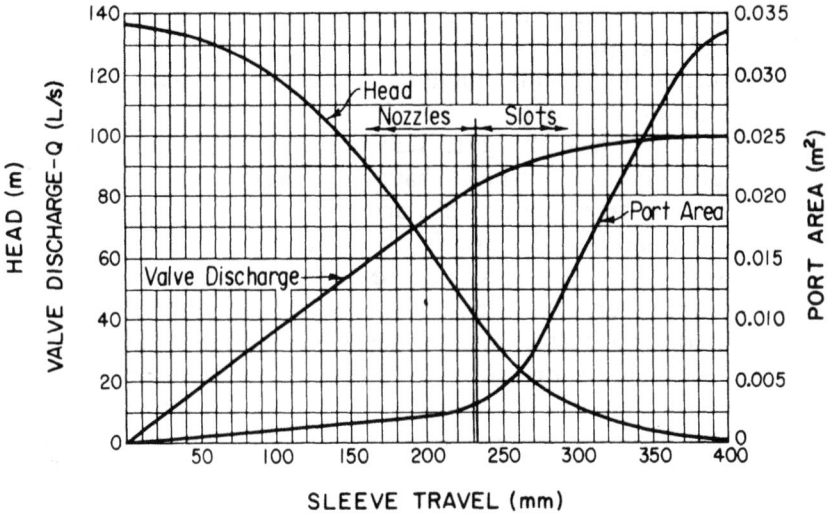

Figure 9-15.–Characteristics of an ideal multijet sleeve valve. 103-D-1734.

9–18. Plug and Ball Valves.–These valves derive their names from the type of control element used–a cone or ball-shaped plug that fits snugly into the valve housing. The plug has a passageway through it that matches the adjacent water passages in the housing when the valve is opened. When rotated 90°, the plug blocks flow through the adjacent passages.

Generally, these valves are not used for regulating because the water passage for partial opening tends to form zones of subatmospheric pressure and cavitation. However, they have been used satisfactorily in combination with orifices to reduce and regulate pressures. Studies of capacity, opening and closing torque, pressure distribution, and head loss should be included in investigations of plug and ball valves.

9–19. Check Valves and Flap Gates.–This type of control is designed to permit flow in one direction only. There are many types of check valves, but the most common has a hinged disk or flap that closes automatically when flow direction is reversed. Another type has a spherical or cylindrical moving element contained within guides. Check valves of this type are usually small. Design of the water passage around the check element is very important. Variations in the flow passage shape resulting from the movement of the element may induce pressure changes that can cause a rapid oscillating motion of the element, and thus produce pulsating flow and vibration. Cavitation is likely to occur for high pipeline

167

velocities if the passage is not designed to eliminate severe subatmospheric pressures.

The outstanding disadvantage of the flap gate when used at the end of a pump discharge line is the intense shock produced by reversed flow when the gate slams shut. Piston-operated and annular-groove water cushions have been designed to reduce the shock resulting from closure of flap gates.

Studies should include capacity, head required to open the gate, and closing characteristics. The relative times at which power interruption, reversal of flow, and closure of gate occurs should be investigated for waterhammer studies. Other factors which must be considered include: pressure intensities at various points in the system, particularly on the pump side of the gate; intensity of the shock, which is extremely difficult to measure and in most cases must be assumed to be proportional to measured movement of the structure or some part thereof; the position of the gate at any instant during the closing cycle; and the angular velocity of the gate at any point in the closing cycle. Electronic instrumentation with an oscillograph is required for recording the events. Records should be taken simultaneously and continuously during the operating cycle. Care must be exercised to assure that some peculiarity of the model does not influence the record which could result in erroneous analysis of the problem.

9–20. Seals.–The primary purpose of the seals described in this section is to prevent leakage past control devices such as gates, valves, or bulkheads. Seals vary widely in shape and materials, and depend on various principles for operation.

Two basic methods are used to effect a seal. The first is by metal-to-metal contact and is used in many small gates and valves where accurate machining is possible and not excessively costly. In this method, direct contact occurs between the moving leaf or vane and the housing or body. As gate or valve size increases, greater clearances are encountered and it becomes necessary to provide a seal capable of filling the larger spaces. The second method then includes sealing the clearance space with a flexible rubber bulb which may be extended hydraulically or mechanically to meet a seal bar or ring on an adjacent member.

Studies of seals differ according to type of control, seal design, expected modes of operation (speed, frequency of closure, etc.), and operating head. Because of difficulties in attaining elastic similitude for model seals and in representing severe subatmospheric pressures that might be present with high velocities, laboratory tests are usually made on sections of the full-size seal under prototype heads. These seal sections and pertinent parts of the gate or control are encased in a housing having transparent windows. The test facilities subject the seal section to conditions that, so far as practicable, are the same as those in the field. Seal behavior and durability under various operating conditions and head, required actuating pressures, seal leakage, wet and dry frictional drag of the seal on the seat, extrusion action of the rubber in clearance spaces, and durability of the seal under reservoir head for long periods are the main characteristics to be considered in a laboratory investigation.

On earlier low-head gate installations, rubber seals having a cross-sectional shape similar to that of a music note were commonly used. When gates subject to higher heads came into use, the music-note seals no longer gave satisfactory service, and a double-stem, bolt-through-type seal was developed. As heads on gates continued to increase, it was necessary to develop a seal that would perform satisfactorily under these higher heads. Double-stem, clamp-on-type seals having 12.7- and 17.5-mm-thick stems were developed, and they proved to have definite advantages over the bolt-through-type seal for high-head applications (fig. 9-16). Variations of the double-stem, clamp-on seals, incorporating fluorocarbon cladding on the sealing surface and brass backing for additional strength, were developed and tested to determine the seal most suitable for very high heads [43, 44].

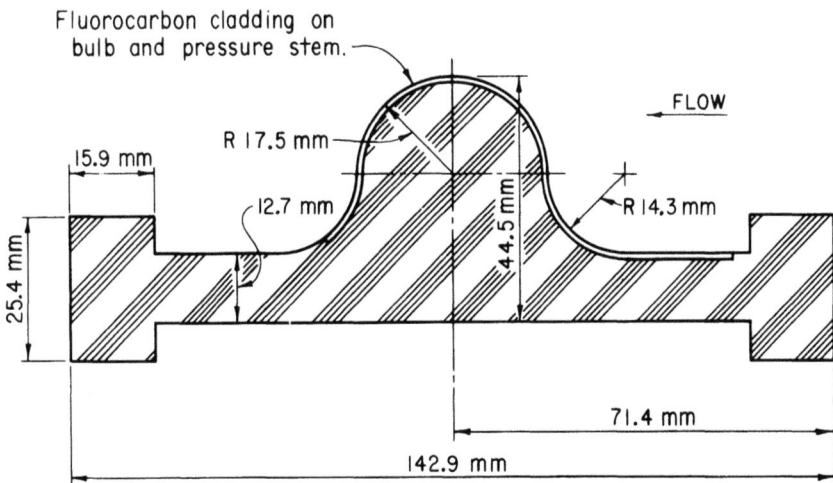

Figure 9-16.–Double-stem, clamp-on-type gate seal. 103-D-1735.

CHAPTER 10

Pumps, Turbines, and
Energy Absorbers

10-1. Pumps.—Mechanical pumps are classified herein according to their principle of operation. When so classified, they fall into three groups:

(1) Centrifugal pumps, utilizing rotation and centrifugal force;

(2) Positive displacement pumps, displacing the fluid by action of rotating or sliding elements in conjunction with check valves; and

(3) Propeller pumps, imparting axial movement to the fluid.

For usual installations, a pump of a given size and efficiency may be selected by referring to data published by pump manufacturers. However, problems sometimes arise in adapting a pump to a given installation. These problems may best be studied in the laboratory, especially where the installation is of unprecedented size, high efficiency of the pump is required, or an unusual operating condition exists. Such tests may be made in advance of installation on a scale model constructed and tested by the manufacturer, or they may be made later on the service pump. Model tests are usually made for large installations, and acceptance of the design is based on the model performance.

Special problems of operation may arise which require a model study. For example, the power required to start a pump may be in excess of the capacity of the power supply line, in which event it is necessary to bring the pump up to speed in air before permitting the water to flow. Studies may be required of the changes in voltage, current, pressure, shock waves, and cavitation, as pumps are started and stopped under various conditions [45]. Entrance elbows and transitions to the pump from a canal or reservoir may be a major part of a model study [46].

10–2. Jet Pumps.–The jet pump (or ejector) is a device that depends on the driving force of a relatively small, high-velocity jet for pumping action (see fig. 10-1). The jet enters a cylindrical chamber of water and imparts motion to the surrounding fluid by an exchange of momentum. The elevation to which the combined flow is lifted depends on the efficiency of the momentum exchange. This type of pump is particularly useful for such purposes as providing cooling water for generators and water for operating fish traps and ladders. Jet pumps of large sizes are often studied in the laboratory prior to fabrication of the prototype [47].

The many variables involved, uncertain theoretical treatment, and the nature of jet-pump action introduce complexities in the design of these pumps. Factors that must be considered in any model study of jet pumps include:

- Shape, velocity, and direction of the driving jet,
- Ratio of diameters of the driving jet and mixing tube,
- Size, length, and shape of mixing tube,
- Size, length, and divergence angle of diffuser tube,
- Size and arrangement of discharge conduit,
- Ease of pumping,
- Quantity of water required,
- Pressure conditions within the pump,
- Possibility of cavitation, and
- Efficiency.

Should some of these factors be incompatible with the others, the efficiency will be affected adversely.

Cavitation can occur at the boundary between the driving jet and the slower moving water set in motion by the jet in the mixing tube. Cavitation pressures in the model may cause difficulty in transferring model results to prototype conditions, requiring special test procedures in which the model is operated at prototype driving, suction, and pumping heads, or a vacuum tank is used to scale the ambient pressure surrounding the pump. When testing in a vacuum tank is not possible, the use of prototype heads is the most reliable method for determining the pressure intensities at points within the model. However, this method does not directly determine the optimum length for the mixing tube or the flow conditions inducing cavitation in the tube.

A jet pump is considered to be well designed when the driving quantity of water is at a minimum for a given driven quantity. The

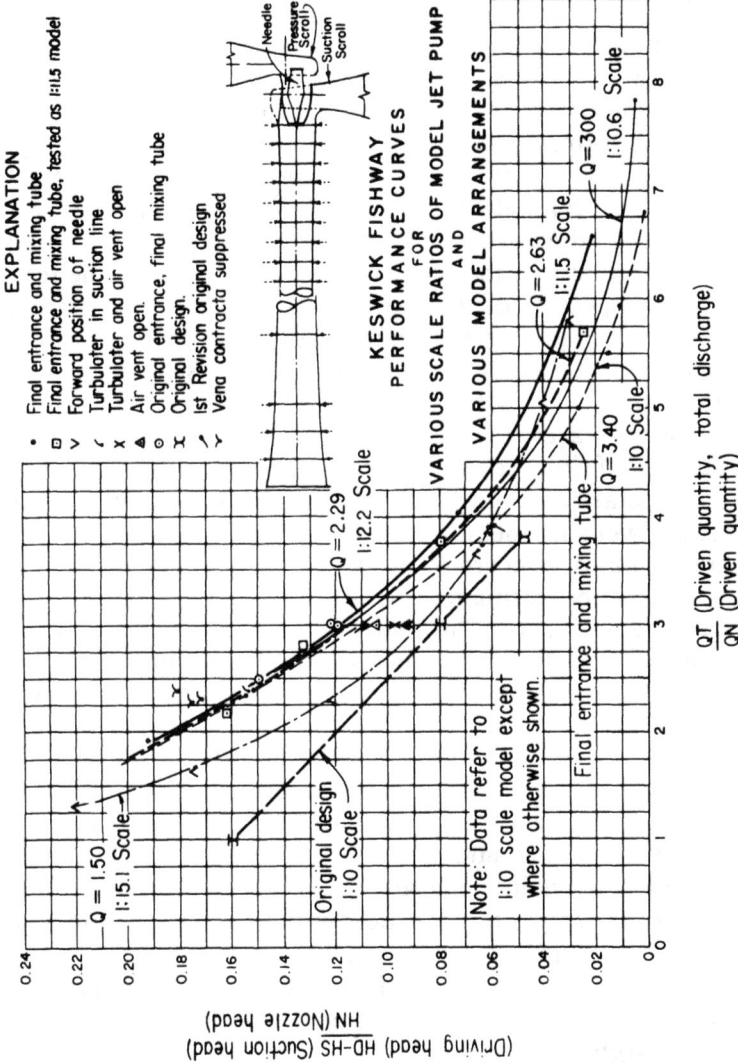

Figure 10-1.–Jet pump cross section and performance curves. 103-D-1736.

173

ratio of these quantities will depend on the ratio of the pumping head to the driving head.

10–3. **Hydraulic Turbines.**–Modern turbines may be classified as impulse, reaction, axial flow, or reversible pump-turbines. Water is conveyed to the turbine by a conduit or penstock and is discharged from the turbine into a draft tube.

Research involved in developing turbine designs, particularly that concerning the turbine spiral case, wicket gates, and runners, has been conducted by turbine manufacturers. Turbines are purchased by performance specifications, in which model studies are required for turbines of unusual size, when unusual operation is expected, or when high efficiency is necessary. The manufacturer constructs and tests the model, and acceptance of the design is based on results of these tests. Laboratory personnel take part in the tests by making an inspection of the model construction and operating procedures and checking the analysis of data, or by actively participating in the model testing and analysis of data. The manufacturer's model may be acquired by the laboratory to study pressure or shock waves resulting from sudden shutdown or change in load, surging of flow, and efficiency of the draft tube. Such tests require a special test stand and involve extensive pressure, discharge, and velocity measurements.

A phenomenon of concern to designers of hydraulic machinery for hydroelectric powerplants is draft-tube surging. Draft-tube surges have been observed in hydroelectric powerplants using Francis-type turbines, apparently since the time these units were first placed in operation. Effects of surges have been observed as power swings or as objectionable noise and vibration. At most powerplants, the operators rapidly learn where the "rough" areas are and avoid operation of the turbines in those regions. At some powerplants, vertical vanes are spaced around the periphery of the draft-tube entrances in an attempt to modify or eliminate the swirl in the flow and thus reduce the severity of, or eliminate, surging. Air is also frequently admitted below the turbine runners to "smooth out" operation.

Surging is known to be caused by rotation of flow passing through the draft tube. A helical vortex is thus generated. Axial flow through a straight tube is stable once the transition to a fully turbulent flow takes place. If rotation is superimposed on this flow, however, the flow pattern makes a drastic change. The axial velocity decreases on the centerline and increases near the wall. The peripheral component of velocity also increases near the wall. Figure 10-2 shows an example of a velocity distribution in axial and

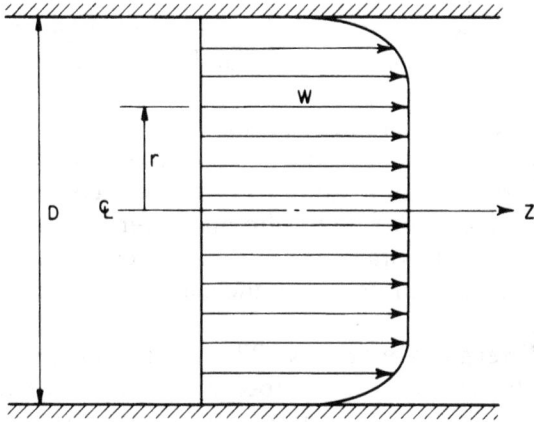

(a) PURELY AXIAL TURBULENT FLOW

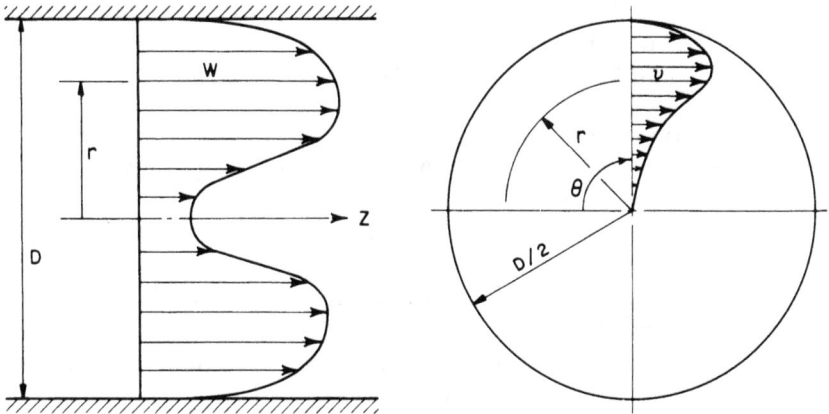

(b) AXIAL FLOW WITH ROTATION

Figure 10-2.–Effect of rotation on flow pattern in turbine draft tubes. 103-D-1737.

175

axial-with-rotation flows. If the discharge is kept constant and the rotational velocity of the flow is increased, a radical change occurs in flow pattern. A reversal in flow takes place at the tube axis and a stagnation point is developed on the centerline. Along the centerline, flow is toward the stagnation point from both the upstream and downstream directions. The development of reversed flow along the axis of the tube has been referred to as vortex breakdown.

Quantitative information regarding both the vortex breakdown and the characteristics of frequency and amplitude of the resultant surge is of considerable value to engineers concerned with the design and selection of hydraulic machinery. Studies have been carried out in the laboratory to produce such quantitative data, as well as qualitative information on the nature of surges in draft tubes.

10-4. Energy Absorbers.–The energy absorber is used to dissipate, absorb, or convert into nondestructive form a large portion of the kinetic energy of a flow bypassing a turbine or powerplant before releasing the water into a natural or constructed channel. An energy absorber for closed-conduit flow makes use of one or more of the following actions:

- Expansion of the flowing jet,
- Impact of the flowing jet on a solid or liquid surface,
- Surface or boundary friction, or
- Disintegration of the flowing jet.

Expansion of the flow to gradually reduce the jet velocity is the most effective of these actions. Most absorber designs include a flow chamber in which the expansion takes place. The shape and size of the chamber depend on local conditions, flow restrictions, and space limitations. Cavitation occurs in energy absorbers of the jet-expansion type when they operate at high heads. Thus, thorough laboratory study of the water passages is necessary. Diffusion of the jet is of prime importance, and the actions listed above are merely means of obtaining the best diffusion.

Extreme care must be taken when conducting model studies of the hydraulics of energy absorbers and in interpreting the resulting data. Because of the energy dissipation within the structure, extensive measurements are necessary during model studies. Fluid turbulence, boundary surface friction, entrainment of air, pressure distribution, and the presence of subatmospheric pressures are important factors that must be investigated.

The pressure measurements should establish turbulence level, air-entrainment characteristics, and the intensity of subatmospheric pressures. Vents for aeration of critical low-pressure areas are effective in minimizing subatmospheric pressures and possible cavitation. Model tests are very useful in establishing the size and location of these vents. Velocity distribution measurements should be made to assure satisfactory performance.

10-5. **Air Diffusers for Ice Prevention.**–In a reservoir, the formation of ice adjacent to trashracks, gates, or other appurtenances on the face of a dam might cause damaging stresses because of expansion and contraction of the ice sheet. Air jets, discharging from nozzles submerged at various locations, have been used successfully to induce circulation of the warmer water from the bottom of the reservoir to the surface, preventing ice from forming near the appurtenances. Many hydraulic problems are involved in the design of such a system. Nozzles for the air jets must give a maximum mixing for a minimum quantity of the circulating fluid (air in this case). The shape of the nozzle passage must not cause excessive expansion of the air, locally or otherwise. Such action would cause both temperature and pressure reductions, often resulting in freezing of the water in the immediate vicinity and plugging of the nozzle. The curve on figure 10-3 shows critical temperature and pressure zones. Hydraulic studies involve measurement of air quantities, pressure differentials, temperatures, effectiveness in inducing circulation, and probability of freezing under operating conditions. An insulated pressure tank may be required for satisfactory control of the water temperature.

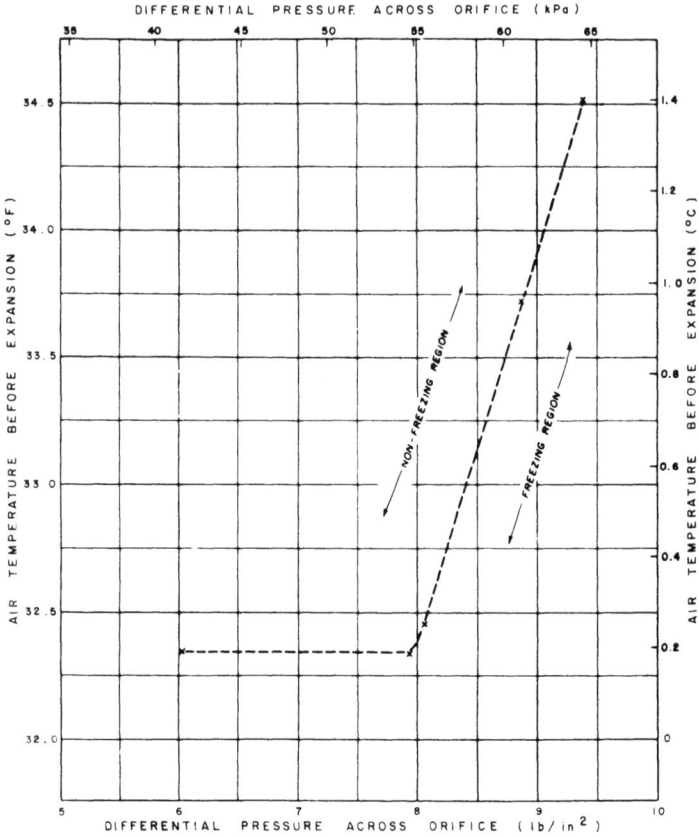

A. CRITICAL FREEZING CURVE FOR ORIFICE "T₂"

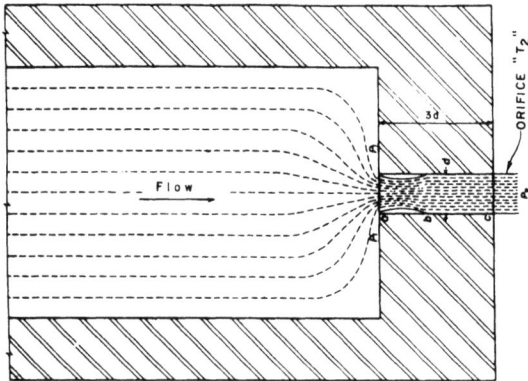

B. FLOW DIAGRAM FOR ORIFICE "T₂"
THROUGH SECTION ON ℄

Figure 10-3.–Typical critical temperature/pressure curve for air
nozzles used in ice prevention systems. 103-D-1738.

CHAPTER 11

Field Studies

A. GENERAL

11-1. Introduction.–Hydraulic field studies are usually considered to be extensions of laboratory investigations to the prototype level. Field and laboratory studies are basically analogous, the primary difference being merely in the size of the structures.

Familiarity with design considerations is necessary in both types of studies. Thorough understanding of the laws of hydraulic similitude, of methods and procedures employed in the laboratory, and of analyses of the results of the investigations is essential in field studies.

For problems of a general nature, a program of laboratory and prototype studies should be planned carefully. In some studies the laboratory cannot provide the solution to a problem without field assistance. An example of such a problem is laboratory simulation of tide. The actual tidal phenomenon must be observed to enable it to be correctly applied to the construction and operation of the model. As another example, laboratory study at model scale would normally be of no value in determining prototype entrainment of air in a high-velocity jet of water.

Field studies do not always require installation and operation of expensive equipment by a large crew of technicians and engineers. Studies can often be performed by one engineer familiar with hydraulic problems who observes a particular action during regular prototype operation. Some investigations, however, do require elaborate equipment and several people to accomplish the task.

Accurate predictions can generally be made from hydraulic model studies. However, certain types of problems can be solved completely only with a full-size structure (see sec. 11–9). When this situation exists, provision should be made for verification of the model test results by prototype investigations. Studies of correlation between model and prototype performances are also valuable in subsequent model and structure design.

Some field studies are performed as a result of operating problems that were not foreseen during the design of the structure. Such studies may wholly or partially solve hydraulic problems and have the advantage of yielding quick results directly from undistorted prototypes, without the expense of constructing and testing a model.

B. MEASUREMENTS

11-2. **General.**–Well-planned and carefully conducted field measurements should have an accuracy comparable to those of laboratory tests. Laboratory equipment may not be usable for field measurements because the forces involved are greater than those encountered in a model. Commercial test equipment is available for some field tests, but it may be necessary to design and build special apparatus for other tests. This does not mean that each particular field test must have specially designed equipment, but, rather, equipment should be designed for repeated use in field applications.

11-3. **Discharge.**–Field installations are not ordinarily equipped with measuring devices intended exclusively for a test program. However, in some cases, permanently installed devices can be used to supplement the field test apparatus. For example, the discharge passing through a structure may be measured at an existing gaging station by use of a current meter or pitot tube, weir, orifice, Parshall flume, or commercial meter. If such devices are not available, consideration should be given to the salt velocity, fluorescent tracer velocity, or the Gibson method. The method selected will depend on the particular conditions, such as size and type of channel or conduit, quantity of water, velocity, and desired accuracy.

11-4. **Pressure.**–Generally, the statements about pressure measurements in Chapter 2, Instrumentation, also apply to prototype studies. Examples of piezometers used for measuring pressure in the field are shown on figures 11-1 and 11-2. Extreme care must be exercised in the installation of piezometers because the velocities encountered in the prototype study are generally higher than in the laboratory. Even a small burr adjacent to the piezometer orifice will seriously affect the readings, and excessive surface roughness, either upstream or downstream from the piezometer tap, will influence the measurements.

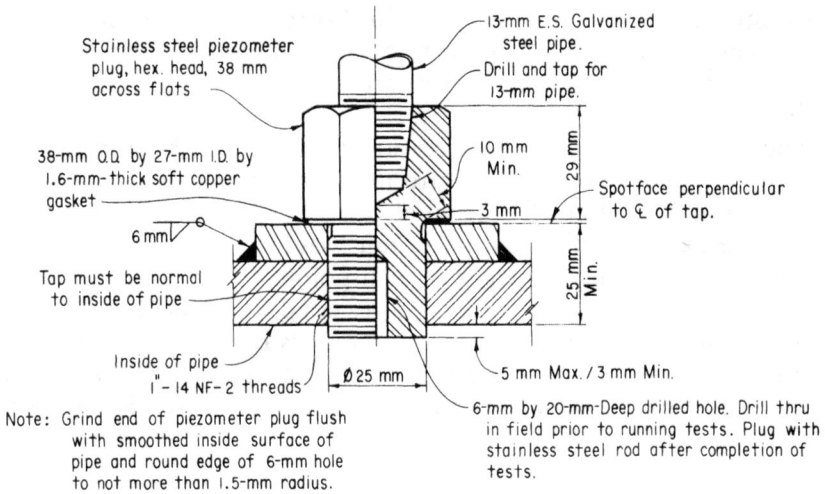

Figure 11-1.–Piezometer connections in steel pipe. 103-D-1739.

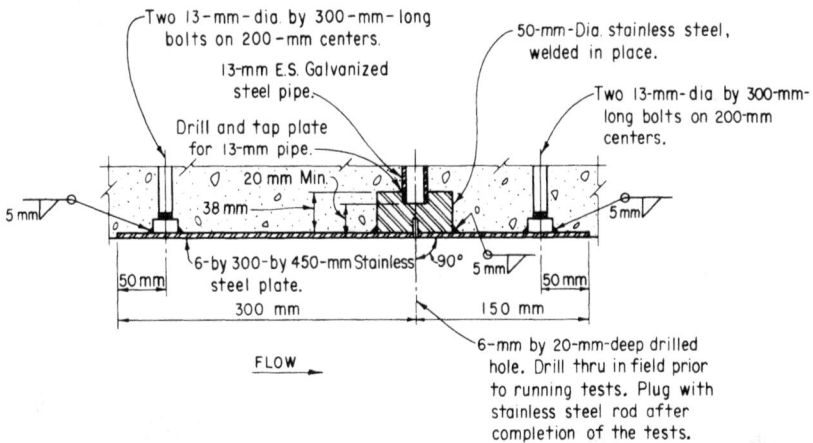

Figure 11-2.–Piezometer installation in concrete. 103-D-1740.

Excluding air from a pressure-measuring system is more difficult in the field than in the laboratory because the prototype flow normally contains more air and the lines are considerably longer. Air entrapment in the lines between the piezometer orifices and the measuring devices should be prevented by locating piezometers away from tops of pipes and other places where air is likely to collect. Accurate pressure readings cannot be obtained if the air is not "bled" from the pressure lines; valves for purging air must be

181

integral parts of the system. A high-pressure water supply, if available, is extremely helpful in purging the system.

Water or mercury manometers are used for low pressures where the length of tubing does not exceed ceiling height or access to the manometer scales. Zero suppression may be used by applying an accurately measured gas pressure through a manifold to the water surface in each tube. Differential pressures between piezometers may be read on tubes shorter than would be necessary for the direct-pressure measurement at each point.

A special type of high-head mercury pot gage has been developed by the Bureau of Reclamation's Hydraulics Branch for use in field studies (fig. 11-3). Gage capacity is equivalent to 90 meters of water. A displacement block and fiducial point are used for accurate

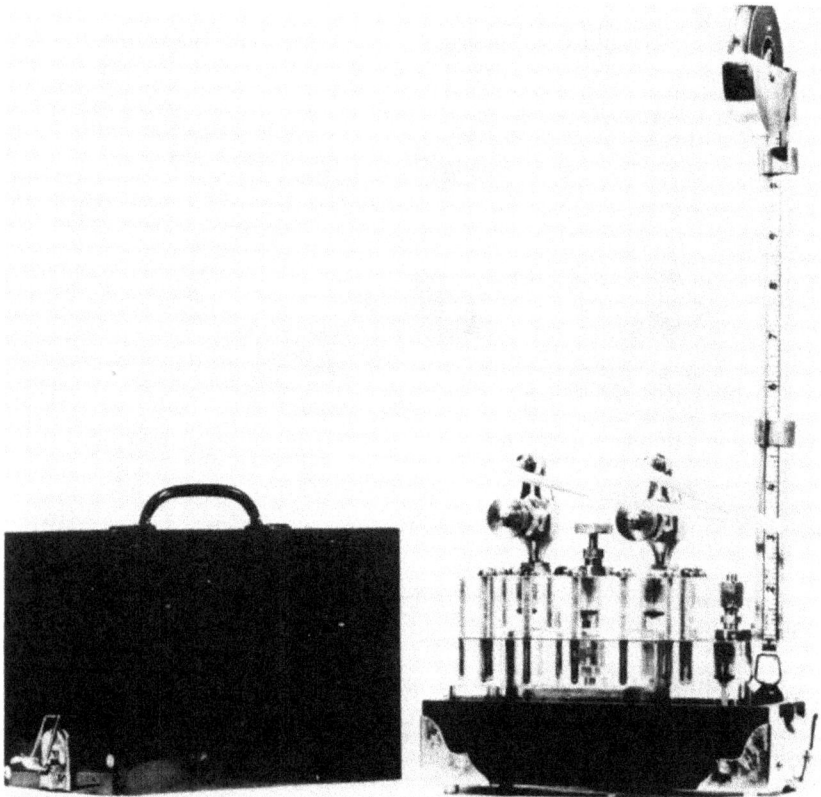

Figure 11-3.–High-head mercury pot gage. P801-D-79213

positioning of the mercury column reference elevation. The gage may be used for direct or differential pressure measurements.

Pressures exceeding the capacity of the mercury gage are measured by a fluid-pressure scale (fig. 2-4B). This instrument has a capacity of 300-pounds-per-square-inch gage pressure (2068 kPa) and is graduated in tenths of a pound per square inch. The scale is readily portable and easy to install and manipulate.

Field studies make use of the recording Bourdon gage. This instrument has the advantage of producing a permanent record, but it must be checked often to ensure accurate results. Clock- or battery-powered pressure-measuring instruments are sometimes necessary for field investigations, particularly in isolated locations. Electrical transducers and recording equipment are used for dynamic measurements in the field. General characteristics of such equipment have been described in chapter 2.

11-5. Water Surface.–Accurate measurements of water surface elevations must be made when investigating freeboard, roughness coefficient, and depth in free-surface structures. Elevations in the laboratory are relatively easy to measure because water surfaces are smooth and well defined in the low-velocity, relatively narrow channels used there. Prototype velocities, on the other hand, may be high and the channels wide; the water surface may be wavy or otherwise poorly defined, or it may be a mass of aerated water.

If prototype conditions are such that the water surface is relatively smooth, or accurate measurements are not required, a simple staff gage mounted on the structure may be used to measure water levels. More accurate data can be obtained by taking measurements in a stilling well which is connected to the water in the structure by a piezometer or pitot tube. One type of gage used in measuring water surface elevations in stilling wells is the electric tape gage. The gage includes a graduated cable or steel tape, weighted at the lower end. The weight and cable (or tape) are part of an electrical circuit that is grounded to the structure. The weight is lowered into the well and a light or meter indicates when it touches the water surface. The instrument is inexpensive and portable. Recording gages utilizing floats are used when data over an extended period of time are desired.

11-6. Velocity.–Velocity measurements are generally made with current meters and various types of pitot tubes. Salt or dye tracers may also be used, most successfully in closed-conduit flows. Velocities in closed conduits can also be measured by inserting reinforced pitot tubes into the flow through circular openings equipped with stuffing boxes. The length of projection into the conduit is limited by the vibration characteristics of the tube in the flowing water [48].

Air velocities are ordinarily measured with pitot tubes or commercial anemometers, and flow quantities are measured by orifices.

11-7. Seepage Loss.–Seepage loss from canals is a problem of growing concern to irrigation engineers. The importance of knowing accurately the magnitude of such losses has increased in recent years as the need for utmost conservation of water and land resources has become more critical.

Laboratory engineers may on occasion be asked to assist in or provide instructions for performing seepage loss measurements in canals or impoundments. These measurements are normally needed for decisionmaking in conserving water, preventing land damage, or in lining the canal or lateral.

Seepage loss measurements may be made on the entire system, on long reaches of a conveyance, or on sections in a canal selected in a manner such that the results will be representative of larger areas. There are several known methods for quantitatively determining seepage losses; the three generally used are: ponding, inflow-outflow, and seepage meter.

Many factors affect the rate of seepage loss from a canal. Some of the more obvious ones are:

- Permeability of material traversed by canal,
- Depth of water,
- Wetted area,
- Location of water table relative to canal invert,
- Slope of subgrade soil structure,
- Flow velocity,
- Soil and water temperatures,
- Entrained air in soil,
- Ground-water inflow,
- Atmospheric pressure,
- Soil and water chemistry, and
- Capillary attraction.

The relative importance of each has not been definitely determined, though it is known that one may offset another and some may even alternately produce an increase and a decrease in seepage rate.

With so many variables operating, equations expressing the interrelationships may never be developed. The theoretical approach to computing seepage losses must be supplanted by measurements. Procedures for performing the measurements are not rigorously defined, but guidelines have been established for the use of devices and methods [49, 50]. These guidelines should be adhered to as closely as possible in both laboratory and field investigations.

11–8. Other Measurements.–Measurement of time is often necessary in field studies. The basic means for time measurement is probably the stopwatch; however, contrary to common opinion, the stopwatch is not a convenient nor neccessarily accurate timer over long periods of time. Where precise measurements are required, a frequency-controlled clock is used with a recorder. In the United States, a timing system may be checked against the highly accurate time signals broadcast by National Bureau of Standards radio station WWV. Broadcast frequencies are 2.5, 5, 10, 15, 20, 25, 30, and 35 megahertz. Station announcements are given in voice and by radio impulses at 5-minute intervals. Another timing device that is acceptably accurate but less complicated than the recording system is a small direct-current motor having a jeweled escapement. The accuracy of this motor is sufficient for short-term field time measurements. Another common method is to use commercial 60-hertz current to drive electric timers. A portable motor-generator is not a dependable source of frequency for timers.

Measurements of length and elevation are very important in field investigations, so known reference points and proven engineering methods should be used. Such measurements should be made relative to established benchmarks and not taken from drawings.

Since gate position indicators are generally not as accurate as desirable, a measurement in field studies usually warranting special consideration is the determination of actual gate positions. Satisfactory measurements may be made by using a leveling rod or flexible steel tape. These measurements may be used in conjunction with the indicator pointer to determine the true gate positions.

Air and water temperatures, relative humidity, barometric pressure, rainfall, and many other conditions that are important in field testing are measured by conventional methods. Recording instruments should be used for investigations covering long time periods to provide data on changes occurring in the absence of an observer.

In the study of vibrations or rapidly fluctuating water pressures, electronic equipment is employed as described in section 2–13. Forces involved in field tests are much greater than encountered in the laboratory, so equipment must be selected to withstand these forces.

185

C. TYPES OF FIELD INVESTIGATIONS

11–9. Introduction.–Hydraulic investigations in the field are generally of the following types:

    (1) Field operating problems,
    (2) Extension of studies to prototype level,
    (3) Model-prototype conformance studies, or
    (4) Performance tests of hydraulic structures and machinery.

There are problems for which complete solutions cannot be obtained in the laboratory. Among these are: (1) negative pressures in conduits, gates, and spillways; (2) coefficients of roughness; (3) vibration; (4) turbulence; (5) high-velocity flow and air entrainment; (6) scour in riverbeds; and (7) air demand. Discussions of all of these are not included in this book. The four main types of studies will be discussed briefly to guide in formulating plans and making decisions. Illustrative examples are included.

11–10. Field Operating Problems.–Occasionally, structures do not operate in the manner conceived by the design engineers, as illustrated in the following example:

> In an outlet works, two 2134-mm-diameter needle valves operating under a maximum head of 49.53 meters discharged directly into a 6100-mm-diameter tunnel. Air was supplied to the tunnel through a 1.83-by 2.44-meter rectangular shaft extending from the top of the dam to the tunnel just downstream from the valves. Low pressure occurred in the valve operating chamber during normal operation, the result being damage to equipment in the service elevator shaft.
>
> The air velocities in the shaft for various valve openings were determined with a pitot tube (fig. 11-4). Measurements at the structure also included pressure in the discharge tunnel, air temperature, barometric pressure, and water discharge. Observations were made of the mechanical behavior of the valves, evidence of cavitation, hydraulic conditions in the stilling pool at the end of the discharge tunnel, and of other associated conditions. The results of these tests permitted a partial solution of the air-demand problem by improving the entrance to the air shaft and thereby minimizing the pressure loss. Basic data were also acquired on air demand of needle valves discharging into tunnels.

This was a study in which very little special equipment was needed.

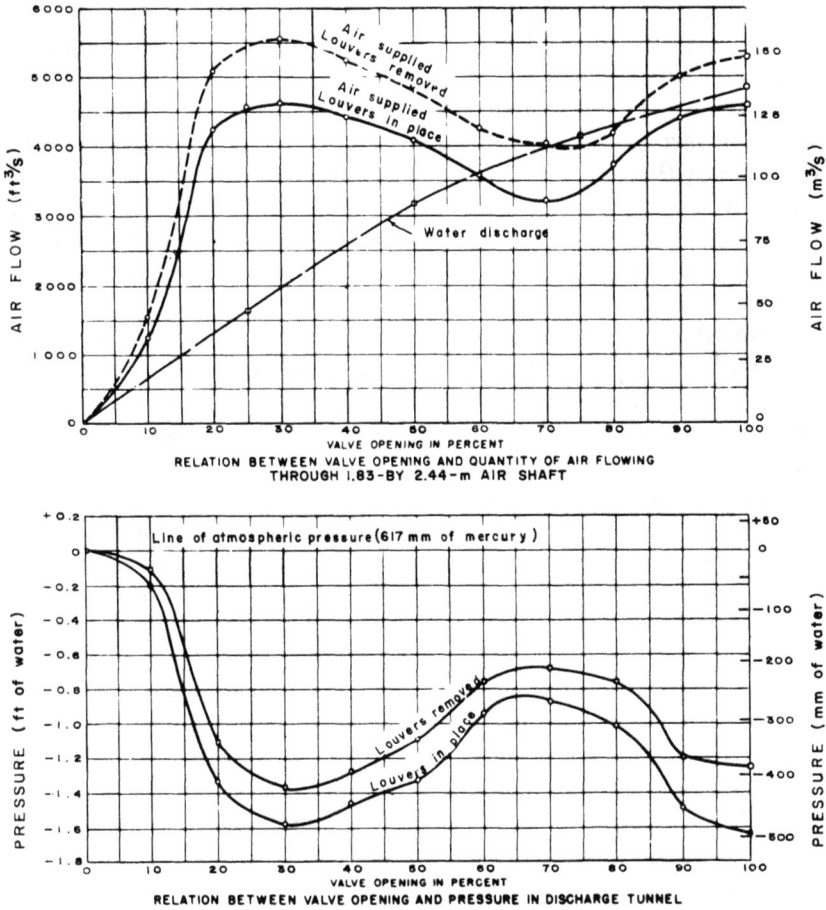

Figure 11-4.–Field performance of needle valve outlet works. 103-D-1741.

11-11. Extension of Model Studies to Prototype Level.–An example of studies that require both laboratory and field work is the investigation of the diffusion of ocean salt water into the estuaries of the Sacramento-San Joaquin River System in California. In this delta, the configuration of the channels and the tidal action cause saline and fresh waters to mix throughout the widths and depths of the flows. These conditions are in contrast to normal action in estuarine channels, where the more dense salt water enters the landward areas as wedges along the bottoms of the watercourses and fresh water flows out to sea over the tops of these wedges.

Development of the agricultural areas in the San Joaquin Valley included two proposed schemes for transport of fresh water across

187

the delta area. One scheme involved construction of a costly artificial channel, and the other contemplated use of natural channels with some artificial cross connection and channel improvement. Hydraulic problems involved in determining whether harmful mixing would occur were very complex examples of unsteady flow. Because theoretical solutions for even the simplest cases of this type are questionable, a hydraulic model was constructed. As the model could not be adjusted perfectly to meet all physical requirements, many of the conclusions were subject to confirmation by prototype tests.

The objectives of the field studies were to: (1) provide information for an orderly step-by-step procedure for construction and overall development of the irrigation project, (2) supply information and practical methods of controlling the intrusion of ocean salinity by releases of stored waters requested by irrigators, and (3) provide information for possible further development in the area. The field tests involved numerous discharge measurements at various points in the delta. These measurements were made with current meters at times when river stages were reflecting seasonal changes in flow. The problem was made more complicated by the tidal action.

An essential part of the study was the measurement of salinity flow direction at various stations to determine salinity gradients. Records from 11 salinity meter stations were averaged, interpreted, and used in conjunction with the distances between the stations to provide values for the salinity gradients. Distribution of saline water in any particular channel was metered by a portable salinity meter mounted on a boat. An electrical analog of the delta was of great assistance in interpreting the field data and in predicting the results of proposed channel changes.

The complexity of salinity intrusion illustrates the fact that a model study may need to be supplemented by prototype study before reaching definite conclusions.

11-12. **Model-Prototype Conformance.**–Although laws governing relationships between a model and its prototype may have been determined, there are similitudes that require confirmation. In these circumstances, field tests of the prototype structure are required for verification of the results from hydraulic model tests.

Such a conformance study was made on a coaster (roller-mounted) gate at Shasta Dam. The gate is designed to close any one of 18 river outlets through the spillway section in the event of a failure of the control valve or for valve maintenance. It is a rectangular steel structure having a skinplate riveted to the downstream side of horizontal beams which are supported by vertical girders and mounted on roller trains (fig. 11-5). In operation, the gate is lowered, by gravity, in guides on the face of the dam.

Normally, a roller-mounted gate is operated under balanced hydrostatic pressure with no flow through an outlet. However,

Figure 11-5.–Coaster (roller-mounted) gate and handling equipment for river outlets. P801-D-79243

design requirements here, and as is the case in most installations, were dictated by the conditions existing during emergency closure under maximum head. Under this condition, the gate would be subjected to large unbalanced pressures. A hydraulic model was used to evaluate the downpull forces. To minimize vibration caused by low pressures in the outlet entrance at partial gate openings, provisions were made in the prototype structure to admit air immediately downstream from the gate. The size of the air supply line was established by the hydraulic model tests.

Field tests were made with a head of 81 meters of water acting on the gate, simulating an emergency condition of operating the gate to close a lower river outlet with the control valve wide open.

(a) *Test Equipment.*–To evaluate the downpull, a resistance-wire strain gage was mounted on the gate stem. The change in resistance of the strain gage was measured by a portable strain indicator. Gate position was indicated by an engineers' tape anchored to the gate and kept taut by a counterweight at the free end. The strain indicator and tape were located close to each other, permitting data to be recorded with a movie camera operating at the rate of approximately three frames per second. A stopwatch was also included in the photographic field to verify the clock in the metering instrument. Simultaneously, the quantity of air admitted immediately downstream from the coaster gate was ascertained by use of a pitot tube. A recording was also made of the minimum pressure in the outlet immediately downstream from the gate.

(b) *Test Procedure and Results.*–Data were acquired on strain, air velocity, pressure, and gate position for both closing and opening cycles. Maximum hydraulic downpull acting on the coaster gate was in reasonable agreement with the value determined from the hydraulic model. The recess in the face of the dam immediately above the entrance to the outlet (fig. 9-5) was effective in equalizing the forces on the upper horizontal seal assembly. However, contrary to model results, this balancing action also occurred at smaller gate openings, producing an uplift force. This force was not large enough to prevent satisfactory movement of the gate during emergency closure.

The prototype investigation showed that model results may not be fully interpreted and thus fail to reveal possible operating difficulties. Prototype and hydraulic model results are compared on figure 11-6.

11–13. **Performance Tests.**–Field performance tests are necessary in many instances to determine the operating

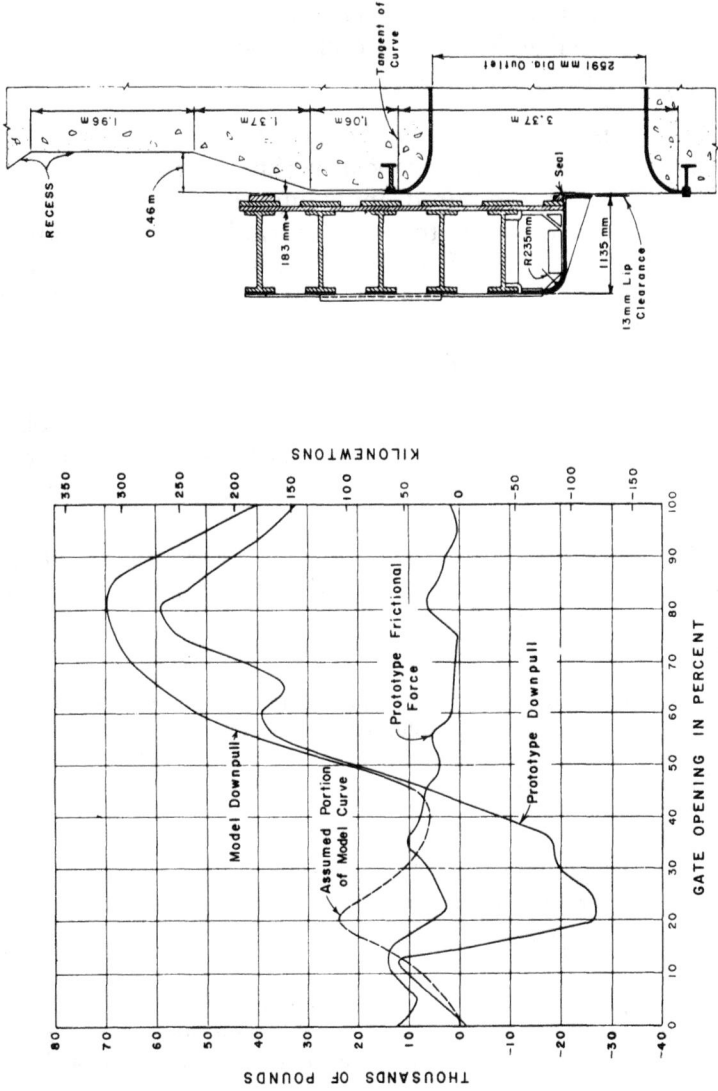

Figure 11-6.–Model-prototype comparison of hydraulic downpull, and prototype frictional force. 103-D-1742.

191

characteristics of hydraulic structures and machines. The reasons for conducting such tests may be to:

- Verify that design criteria have been met,
- Provide a basis for settlement between the contractor and the purchaser,
- Acquire data for future designs, or
- Determine if additions or modifications are necessary in the structure or machine.

Extensive tests can be made on the model of a structure, but the results of the studies may be inconclusive; short- or long-term field investigations may be required, involving visual observations of prototype operation or extensive hydraulic and mechanical measurements. Inspection of prototype structures may be necessary to disclose changes in solid boundary surfaces, erodible channels, and structural parts. Corrective measures may be suggested immediately following completion of the tests. By combining model and prototype tests, prototype operating difficulties may be minimized and greater confidence gained in the use of models.

Field performance tests on hydraulic machines include those on turbines and pumps. Procedures and techniques for testing hydraulic machinery appear in the Power Test Codes for Hydraulic Prime Movers, published by the American Society of Mechanical Engineers, and in standards published by international organizations.

The description of a typical prototype test follows:

> A pump test was conducted to establish as much of the characteristic performance curves as possible. The test was made at the head available between intake and discharge water surface levels. Additional discharge heads were obtained by partially closing a temporary gate that controlled the elevation of the water surface in a tower at the end of discharge lines.
>
> Data required for determining pump performance include:
>
> - Power input to the motor,
> - Total pumping head, and
> - Pump discharge rate.
>
> Additional observations should be made of: temperatures of the water, motor room air, air discharged from the motor windings, motor core iron, and motor thrust bearing; vibrations; and line voltage and amperage and exciter amperage.
>
> Power input to the motors was measured by two calibrated watt-hour meters. The suction head was measured with a mercury manometer, and the discharge pumping head with a 7.62-meter mercury gage. Discharge from the pumps was measured over a standard weir. Results of the tests of one unit are shown on figure 11-7.

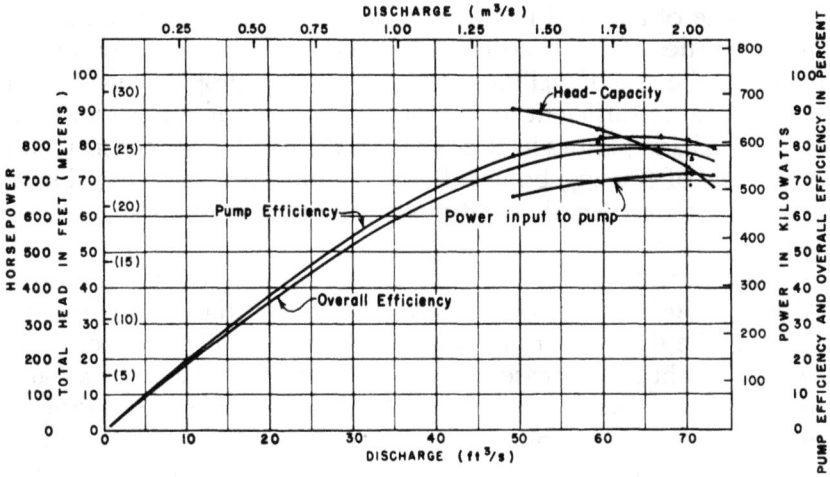

Figure 11-7.–Pump performance curves. 103-D-1743.

11-14. Floating Equipment.–Floating equipment is often used in the construction, operation, and maintenance of dams and their appurtenant features, but hydraulic laboratory testing is seldom involved. Occasionally, there is need for special equipment such as caissons and bulkheads to allow unwatering for inspection and repair of underwater facilities or portions of a structure. The properties, characteristics, and design criteria applying to these structures are the same as those used in designing boats, ships, drydocks, and other floating structures. Hydraulic model tests may include determination of:

- Size and direction of forces to be exerted on cables used for maneuvering facilities;
- Coefficient of drag for various positions of the equipment for different velocities of flow;
- Maneuvering procedures;
- Stability or ability of equipment to remain upright or within a given angle of tilt or list during operation;
- Amount and location of ballast effecting stability;
- Effect of metacentric height on motion of the equipment in choppy waters;
- Magnitude and effect of velocity and turbulence on equipment in position on the structure;
- Procedures to be used in flooding and unwatering working compartments; and
- Hydraulic characteristics of auxiliary equipment such as drydocks, special operating barges, or special floating compartments.

193

The size of a model for flotation studies is usually governed by the properties of the material to be used in the construction of the model or the size of facilities available for model operation. Model size should be sufficient to minimize the effect of surface tension. Both metal and plastics have been used successfully for construction of models of floating bodies. Plastics should be types that absorb very little water. An all-metal model of a floating caisson is shown in figure 11-8.

Forces acting on towed or maneuvered floating equipment in flowing water result from relative motions of the vessel and the water. If the flow pattern is complicated by several sources of flow, such as might be the case immediately downstream from a dam, model studies may be necessary in establishing the number of maneuvering lines required and determining the orientation of these lines with respect to the path of the vessel, measuring the forces acting on each line (for the purpose of establishing cable size), and designing the capacities of operating hoists, puller machines, or other towing facilities.

A. Side view of caisson, powered operating barge in lowered position. P801-D-79241

B. Oblique view of caisson showing working space. Operating barge in raised position. P801-D-79242

Figure 11-8.–Complete 1:20 scale model of the floating caisson and operating barge used in the repair and maintenance of the spillway bucket at Grand Coulee Dam.

195

BIBLIOGRAPHY

[1] *Fluid Meters, Their Theory and Application*, sixth ed., American Society of Mechanical Engineers, pp. 197-230, 1971.

[2] Marks, Lionel S., "Square-Edged Inlet and Discharge Orifices for Measuring Air Volumes in the Testing of Fans and Blowers," *Transactions of the American Society of Mechanical Engineers*, vol. 58, AER-58-7, p. 593, 1936.

[3] O'Brien, M. P., and R. G. Folson, "Modified I.S.A. Orifice with Free Discharge," *Transactions of the American Society of Mechanical Engineers*, vol. 59, RP-59-1, p. 61, 1937.

[4] Lansford, W. M., *The Use of an Elbow in a Pipeline for Determining the Rate of Flow in the Pipe*, Bulletin No. 289, Engineering Experiment Station, University of Illinois, Urbana, Ill., 1936.

[5] *Water Measurement Manual*, second ed., rev. reprint, Bureau of Reclamation, 1974.

[6] *Pressure Handbook*, Rimbach Publications, Division of Chilton Company, Pittsburg, Pa., 1967.

[7] Cerni, R. H., and L. E. Foster, *Instrumentation for Engineering Measurement*, third printing, John Wiley & Sons, Inc., 1966.

[8] Lion, Kurt S., *Instrumentation in Scientific Research*, McGraw-Hill, 1959.

[9] Ramey, Robert L., *Electronics and Instrumentation*, Wadsworth Publishing Co., Inc., 1963.

[10] Buckingham, E., "On Physically Similar Systems; Illustrations of the Use of Dimensional Equations," *Physical Review*, vol. 4, p. 345, 1914.

[11] Albertson, M. L., J. R. Barton, and D. B. Simons, *Fluid Mechanics for Engineers*, Prentice-Hall Civil Engineering and Engineering Mechanics Series, p. 21, 1960.

[12] Rouse, H. (editor), J. E. Warnock, "Hydraulic Similitude," *Engineering Hydraulics*, John Wiley & Sons, Inc., p. 136, 1950.

[13] Kline, S. J., *Similitude and Approximation Theory*, McGraw-Hill Book Co., Inc., 1965.

[14] Maxwell, W. H. C., and J. R. Weggel, "Surface Tension in Froude Models," *Journal of the Hydraulics Division*, Proceedings of the American Society of Civil Engineers, Paper 6482, vol. 95, No. HY2, p. 677, March 1969.

[15] Craya, A., "Similitude des Modeles Fluviaux a Fond Fixe," *La Houille Blanche*, vol. 3, No. 4., p. 346, 1948.

[16] "Studies of Crests for Overfall Dams," *Boulder Canyon Project Final Reports, Part VI–Hydraulic Investigations*, Bulletin No. 3, Bureau of Reclamation, 1948.

[17] Peterka, A. J., *Hydraulic Design of Stilling Basins and Energy Dissipators*, Engineering Monograph No. 25, rev. ed., Bureau of Reclamation, 1963.

[18] Bowers, C. E., and F. Y. Tsai, "Fluctuating Pressures in Spillway Stilling Basins," *Journal of the Hydraulics Division*, Proceedings of the American Society of Civil Engineers, vol. 95, No. HY.6, p. 2071, November 1969.

[19] Beichley, G. L., *Hydraulic Design of Stilling Basin for Pipe or Channel Outlets*, Research Report No. 24, rev. ed., Bureau of Reclamation, 1976.

[20] Simmons, W. P., Jr., *Hydraulic Design of Transitions for Small Canals*, Engineering Monograph No. 33, Bureau of Reclamation, 1964.

[21] Einstein, H. A., and N. Chien, "Similarity of Distorted River Models With Movable Bed," *Transactions of the American Society of Civil Engineers*, Paper No. 2805, vol. 121, p. 440, 1956.

[22] Martin, H. M., and E. J. Carlson, "Model Studies of Sediment Control Structures on Diversion Dams," *Proceedings, Minnesota International Hydraulics Convention*, International Association for Hydraulic Research, American Society of Civil Engineers, September 1-4, 1954.

[23] Carlson, E. J., *Removal of Saline Water From Aquifers*, Research Report No. 13, Bureau of Reclamation, 1968.

[24] Brooks, N. H., and R. C. Y. Koh, "Selective Withdrawal from Density-Stratified Reservoirs," *Journal of the Hydraulics Division*, Proceedings of the American Society of Civil Engineers, vol. 95, No. HY4, p. 1369, July 1969.

[25] King, D. L., *Hydraulics of Stratified Flow, Second Progress Report, Selective Withdrawal From Reservoirs*, Report No. HYD-595, Bureau of Reclamation, September 1969.

[26] King, D. L., *Reaeration of Streams and Reservoirs–Analysis and Bibliography*, Report No. REC-OCE-70-55, Bureau of Reclamation, December 1970.

[27] Palde, U. J., *Hydraulic Laboratory Studies of a 4-Foot-Wide Weir Box Turnout Structure for Irrigation Use*, Report No. REC-ERC-72-31, Bureau of Reclamation, September 1972.

[28] Knapp, R. T., J. W. Daily, and F. G. Hammitt, *Cavitation*, Engineering Societies Monograph, McGraw-Hill Book Co., 1970.

[29] Bradley, J. N., and L. R. Thompson, *Friction Factors for Large Conduits Flowing Full*, Engineering Monograph No. 7, rev. reprint, Bureau of Reclamation, 1977.

[30] Cassidy, J. J., *Control of Surging in Low-Pressure Pipelines*, Report No. REC-ERC-72-28, Bureau of Reclamation, November 1972.

[31] Streeter, V. L., and E. B. Wylie, *Hydraulic Transients*, McGraw-Hill Book Co., 1967.

[32] McBirney, W. B., "Some Experiments with Emergency Siphon Spillways," *Journal of the Hydraulics Division*, Proceedings of the American Society of Civil Engineers, vol. 84, No. HY5, p. 1807, October 1958.

[33] Kalinske, A. A., and James M. Robertson, "Entrainment of Air in Flowing Water–Closed Conduit Flow," *Transactions of the American Society of Civil Engineers*, Paper No. 2205, vol. 108, p. 1435, 1943.

[34] Babb, A. F., J. Amorocho, and J. Dean, *Flow Behavior in Large Siphons*, Paper 1016, Department of Water Science and Engineering, University of California, Davis, Calif., February 1967.

[35] Tessitor, F., and D. J. Hebert, *Hydraulic Model Studies of a Siphon Elbow Proposed for the Grand Coulee Pumping Plant Discharge Lines*, Report No. HYD-163, Bureau of Reclamation, March 1945.

[36] Kotz, S. E., and W. P. Simmons, Jr., *Hydraulic Model Studies of the Siphon and Feeder Canal Transition for Grand Coulee Pumping Plant*, Report No. HYD-224, Bureau of Reclamation, March 1949.

[37] *Entrances to Conduits of Rectangular Cross Section*, Technical Memorandum No. 2-428; *Investigation of Entrance Flared in Four Directions*, Report No. 1, March 1956; *Investigation of Entrances Flared in Three Directions and in One Direction*, Report No. 2, June 1959, Corps of Engineers, U.S. Army Engineers Waterways Experiment Station, Vicksburg, Miss.

[38] Falvey, H. T., "Hydrodynamic Pressures in Conduits Downstream from Regulating Gates," *Proceedings of the 12th Congress of the International Association for Hydraulic Research*, 1967.

[39] Davis, C. V., and K. E. Sorenson, *Handbook of Applied Hydraulics*, third ed., Section 22, McGraw-Hill Book Co., 1969.

[40] Warnock, J. E., and H. J. Pound, "Coaster Gate and Handling Equipment for River Outlet Conduits in Shasta Dam," *Transactions of the American Society of Mechanical Engineers*, vol. 68, No. 3, p. 199, 1946.

[41] Dodge, R. A., *Hydraulic Downpull Studies of the Fixed-Wheel Spillway Gates for Red Bluff Diversion Dam, Central Valley Project, California*, Report No. HYD-511, Bureau of Reclamation, April 1963.

[42] Hebert, D. J., and J. W. Ball, "The Development of High Head Outlet Valves," *Report on Second Meeting*, Appendix 14, International Association for Hydraulic Structures Research, Stockholm, Sweden, June 6-7, 1948.

[43] Mohrbacher, R. D., *High Head Gate Seal Studies*, Report No. HYD-582, Bureau of Reclamation, March 1968.

[44] Traut, E. J., *High Head Gate Seal Studies*, Report No. REC-OCE-70-37, Bureau of Reclamation, August 1970.

[45] Bradley, J. N., *Studies to Determine Suitable Methods for Starting and Stopping the Pumps in the Granby Pumping Plant–Colorado-Big Thompson Project, Colorado*, Reports No. HYD-113 and HYD-150, Bureau of Reclamation, June 1942 and September 1944, respectively.

[46] Isbester, T. J., *Model Studies of Suction Tubes for Mile 18 and Forebay Pumping Plants, San Luis Unit, Central Valley Project, California*, Report No. HYD-513, Bureau of Reclamation, June 1963.

[47] Locher, Fred, *Hydraulic Studies of a Water Jet Pump for the Keswick Dam Fishtrap–Central Valley Project, California*, Report No. HYD-154, Bureau of Reclamation, September 1944.

[48] Winternitz, F. A. L., "Effects of Vibration on Pitot Probe Readings," *The Engineer*, Parts 1 and 2, vol. 201, Nos. 5227 and 5228, pp. 273 and 288, March 30 and April 6, 1956.

[49] Robinson, A. R., and Carl Rohwer, *Measuring Seepage from Irrigation Canals*, Technical Bulletin No. 1203, Agricultural Research Service, Colorado Agricultural Experiment Station, Fort Collins, Colo., 1959.

[50] McBirney, W. B., *Measuring Seepage Loss in Irrigation Canals*, Report No. HYD-459, Bureau of Reclamation, March 1961.

INDEX

☆ U.S. Government Printing Office: 1986—652-671/227

Mission of the Bureau of Reclamation

The Bureau of Reclamation of the U.S. Department of the Interior is responsible for the development and conservation of the Nation's water resources in the Western United States.

The Bureau's original purpose "to provide for the reclamation of arid and semiarid lands in the West" today covers a wide range of interrelated functions. These include providing municipal and industrial water supplies; hydroelectric power generation; irrigation water for agriculture; water quality improvement; flood control; river navigation; river regulation and control; fish and wildlife enhancement; outdoor recreation; and research on water-related design, construction, materials, atmospheric management, and wind and solar power.

Bureau programs most frequently are the result of close cooperation with the U.S. Congress, other Federal agencies, States, local governments, academic institutions, water-user organizations, and other concerned groups.

www.ingramcontent.com/pod-product-compliance
Lightning Source LLC
Chambersburg PA
CBHW031950180326
41458CB00006B/1679